D0641055

Estimation and Mental Computation

1986 Yearbook

Harold L. Schoen
1986 Yearbook Editor
University of Iowa

Marilyn J. Zweng
General Yearbook Editor
University of Iowa

National Council of
Teachers of Mathematics

Library of Congress Cataloging in Publication Data:

Main entry under title:

Estimation and mental computation.

(Yearbook / National Council of Teachers of Mathematics ; 1986)
Bibliography: p.
1. Problem solving. 2. Estimation theory. 3. Arithmetic, Mental. I. Schoen, Harold L. II. Zweng, Marilyn, 1932– III. Series: Yearbook (National Council of Teachers of Mathematics) ; 1986.
QA1N3 1986 [QA63] 510s [511'.4] 85-31947
ISBN 0-87353-226-0
ISSN 0077-4103

The publications of the National Council of Teachers of Mathematics present a variety of viewpoints. The views expressed or implied in this publication, unless otherwise noted, should not be interpreted as official positions of the Council.

Printed in the United States of America

Contents

PART 4: TEACHING MEASUREMENT ESTIMATION

PART 5: TESTING AND RESEARCH

Preface

The 1986 Yearbook is a response to the growing recognition among mathematics educators of the importance of teaching and learning estimation. The only previous yearbook to focus on any aspect of estimation was *Approximate Computation* (1937), which dealt primarily with the mathematical theory of approximate numbers. In 1986, partially because of advances in calculator and computer technology, a broader view of estimation—along with its companion, mental computation—is required. Hence, this yearbook makes a strong case for the importance of estimation in various domains, for example, numerosity, computation, and measurement.

One important theme that arises frequently throughout the volume is the interdependency between the process of estimating in a particular domain and the understanding of mathematical concepts within that domain. Estimating aids in concept development, but at the same time a solid conceptual understanding improves one's ability to make good estimates. It follows as a corollary to this theme that an important by-product of learning to estimate is better conceptual understanding, and, conversely, concepts must be understood in order to acquire the flexible set of processes and decision-making rules needed by the proficient estimator.

The yearbook is divided into five sections, with a short "Benchmark" between sections. The Benchmarks contain topical teaching ideas or interesting information concerning estimation or mental computation in the areas that are described in more detail in the subsequent sections.

The first section (articles 1–5) provides the framework for the yearbook. Article 1 describes not only a wealth of situations in which estimation is useful but also many others where estimation is, in fact, the only reasonable approach. With the need for estimation established, articles 2 and 3 discuss the thinking strategies and mind-set necessary to be a good computational estimator and offer many ideas for helping students develop them. Article 4 turns to mental computation, discussing some of its history in the schools and establishing a modern-day need for mental computation as a corequisite for computational estimation. Finally, article 5 deals in a fairly nontechnical way with the arithmetic of approximate numbers.

Part 2 (articles 6–14) presents specific instructional activities. It contains programs of study for teaching mental computation and different types of estimation in the primary grades and in the crucial areas of fractions, decimals, and percents.

Part 3 (articles 15–21) stresses the remarkable variety of types of estimation and uses for estimation. These articles describe the role of estimation in a teacher-parent workshop, in a game for high school students, in a calculator-aided activity, in writing a computer program, and in solving a mathematical problem.

Part 4 (articles 22–25) focuses on just one important type—estimation in measurement—including teaching activities and mathematical topics appropriate for grades K–12. Finally, Part 5 (articles 26 and 27) comprises a review of research on teaching and learning estimation and a discussion of some of the difficulties and recent progress in attempts to test students' ability to estimate.

Many people deserve credit for making this yearbook possible. From the Educational Materials Committee, which decided several years ago that the 1986 Yearbook would address the topic of estimation, to the highly competent NCTM Reston staff, led by Charles Hucka, who were responsible for its final production, this has been a team effort. Special thanks to the authors whose contributions of time, effort, and expertise have made this volume the definitive work on estimation in mathematics education at this time. Special thanks, too, to Eric Hart, my graduate assistant, for his insightful comments and careful content editing. Finally, I am very grateful to the review committee—especially my colleague and the general yearbook editor, Marilyn Zweng—who lent their talent and expertise at the crucial, formative stages of this yearbook. My sincere thanks to each of these committee members:

Cheryl Arevalo	Des Moines Schools, Des Moines, Iowa
David Duncan	University of Northern Iowa, Cedar Falls, Iowa
Barbara Reys	University of Missouri, Columbia, Missouri
Paul Trafton	National College of Education, Evanston, Illinois
Fred Weaver	University of Wisconsin, Madison, Wisconsin
Marilyn Zweng	University of Iowa, Iowa City, Iowa

As my work on this yearbook progressed, my appreciation of the importance of estimation in mathematics education deepened. For me it was an exciting, stimulating experience. I hope and fully believe that you will enjoy an equally rewarding experience.

HAROLD L. SCHOEN
1986 Yearbook Editor

1

Reasons for Estimating

Zalman Usiskin

MATHEMATICAL ideas survive over a long period of time only if they are wide in applicability. The use of estimation dates back at least as far as ancient attempts to measure land area and time. Mathematical ideas involving estimation began then as well and have occupied many of the most famous mathematicians of all times. For instance, over two thousand years ago, Archimedes estimated pi as $223/71 < \pi < 22/7$, and in the seventeenth century, Newton developed a sophisticated method for approximating solutions to equations. Within the past hundred years we have witnessed the emergence of statistics as an important discipline that studies populations through estimates based on sampling. Yet even today many people view estimation as somehow foreign to the mainstream of mathematics.

This article begins with an examination of the reasons why estimation is viewed as foreign. Then, after some discussion of ideas relating to estimates, four major reasons for doing estimating are described and exemplified. Finally, reasons for teaching estimation are offered.

ESTIMATION IN MATHEMATICS

Mathematics is commonly depicted as involving only the most exact of ideas:

> What is exact about mathematics but exactness.—Goethe
>
> Mathematics, the priestess of definiteness and clearness.—J. F. Herbart

This exactness is felt to have implications beyond mathematics itself:

> Numerical precision is the very soul of science.—D'Arcy Wentworth Thompson
>
> The advancement and perfection of mathematics are intimately connected with the prosperity of the State.—Napoleon Bonaparte

---

This article is an adaptation of chapter 12, "Reasons for Estimating," of *Applying Arithmetic,* by Zalman Usiskin and Max Bell. That book was developed with the assistance of grant 79-19065 of the National Science Foundation. I wish to thank Max Bell for helpful suggestions on a draft. The sole responsibility for the opinions presented here lies with the author.

This precision is particularly considered applicable to the reasoning found in mathematics:

The reasoning of mathematics is a type of perfect reasoning.—P. A. Barnett

Estimation does not seem to fit in all this. The *Random House College Dictionary* definition of the verb *estimate* is "to form an approximate judgment or opinion regarding the value, amount, size, weight, etc., of; calculate approximately." This definition does not match well with "perfect reasoning" or "numerical precision" and seems the very opposite of exactness. As a noun, the definition is no better: an *estimate* is "an approximate judgment or calculation, as of the value, amount, or weight of something." The word *estimation* is likewise used to represent both the process and the result of estimating, and the phrase *in my estimation* signals a value judgment, not somthing one expects in mathematics. The idea of *approximation* fares no better. The verbs *approximate* and *estimate* are synonyms.

Occasionally semantics confuses the issue. Consider the common estimate 3.14 for π. As a number, 3.14 is as exact as π, and calculations with 3.14 are as precise as calculations with π. Yet calculations with 3.14 are often called calculations with "approximate numbers," and in accordance with the dictionary definition above, one may be asked to use the estimate 3.14 to "calculate approximately" the area of a circle. The phrases *calculate approximately* and *approximate number* are each self-contradictory, what rhetoricians term oxymorons.

Thus both in concept and in terminology estimation seems antithetical to the very nature and purposes of mathematics. Indeed, estimating is often viewed as a task one engages in as a less attractive alternative to dealing with exact information.

But an analysis of the reasons for estimating verifies just the opposite—estimating is often more reasonable than avoiding estimates, and estimating is often the only choice one has in a situation. Furthermore, the uses of estimation fit the ideals of mathematics, namely, clarity in thinking and discourse, facility in dealing with problems, and consistency in the application of procedures.

Before we detail the reasons for estimating, it helps to clarify a few mathematical ideas about estimates. As a *mathematical object,* an estimate is usually either a single number or an interval (a set of numbers between two given numbers). Examples of the latter are Archimedes' estimate for π and the prediction that tomorrow's winds will be from 15 to 20 mph. Estimates are intervals more often than one would expect from textbook examples. Geometric estimates are usually connected regions or neighborhoods. For instance, we may estimate the location of the epicenter of an earthquake to be within 10 km of a particular point.

As a *mathematical process,* estimating may be considered to fall between rewriting, that process in which a number is replaced by its equal, and transforming, in which a number is replaced by its corresponding value in a function. For example, a score of 41 right out of 50 on a standardized test may be written as 41/50 and recorded as 82% (rewriting) or found to be at a grade level of 6.1 (transforming), which we say is about 6th grade level (estimating). The numbers 41/50 and 82% are equal; 6.1 and 6th grade level are close; and 82% and 6.1, though associated with each other, are numerically quite far apart.

The broad *mathematical context* for an estimate is usually one of the following types:

A. An exact value is known but for some reason an estimate is used (e.g., the estimate 1.732 for $\sqrt{3}$).
B. An exact value is possible but is not known and an estimate is used (e.g., the age of an old sequoia tree before it is chopped down).
C. An exact value is impossible (e.g., the estimated life of a light bulb).

At one time Max Bell and I tried to sort out the semantics of these contexts. We knew that the word *approximation* is often used in situations of type A and *estimate* is often used for types B and C. However, an examination of books and dictionaries convinced us that in common usage the words are interchangeable. Any attempt to define a difference between these words might not be applicable in situations outside the definer's immediate environs. Hence no distinction between approximation and estimation is attempted in this article.

The *mathematical procedures* used in estimation range from making ball-park estimates, or guesstimates, through the various kinds of rounding (up, down, or to the nearest multiple of a given number), to sophisticated procedures like those used in mathematical analysis, statistics, or economics. Though this article emphasizes number estimates, one may estimate geometric figures, functions, and many other mathematical objects. Obviously the procedure used depends on the type of estimate desired, the mathematical or real context, and the knowledge of the person doing the estimating. As the other articles in this volume illustrate, a wide variety of procedures are possible.

The wealth of procedures in use results from the many reasons and contexts in which estimation is used. We now examine these reasons and illustrate each reason with a variety of contexts.

REASON 1: CONSTRAINTS FORCE ESTIMATES

In a large number of situations, there is no choice but to estimate, because exact values are not obtainable. Here is a brief typology of such constraints:

(a) Value unknown. A value may be unknown, forcing an estimate. Predictions of the future, guesses about the past, estimates of military

strength or economic conditions in countries, and even educated guesses regarding what groceries will cost are estimates forced by lack of knowledge.

(b) Value varies. A value may be different each time it is measured, forcing one to estimate it. Temperatures, populations, air pressures, and typing speeds are common examples. Situations involving relative frequencies, such as the number of heads in 100 tosses of a coin, are of this type.

(c) Measure limitations. Physical measurements are, for the most part, not exact. Objects do not have precise lengths (look under a microscope and the edge of this paper is seen as rough) and measuring devices have imperfections. Thus many measures that seem exact are more accurately viewed as estimates close enough for practical considerations.

Reasons *(a)*, *(b)*, and *(c)* are reasons for *sampling* in statistics. This is no coincidence; a sample is an estimate of an entire population. Examples 3, 8, 14, and 16 below illustrate that there are more connections between estimation and statistics than textbooks suggest.

(d) Limited domain. A value may make sense only as a whole number or member of some other fixed set, so any other result must be adjusted. For instance, if candy is priced at 3 for 10¢, because there are no coins for parts of a penny, a person will pay 4¢ for a single piece. A different type of limit on the domain is illustrated in the following example: If a new highway is to be built and the original route is unsuitable, a route approximating the original will have to be found.

(e) Safety margins. A margin for error may dictate an estimate. The capacity of an elevator must be an underestimate of what the elevator can actually hold. It may be desirable for a shopper going to a store to overestimate both the time it will take (so as to leave enough time to buy) and the amount it will cost (so as to have enough money to cover the purchase).

(f) Estimate in, estimate out. A value may itself have been calculated from estimates, so anything calculated from it must be an estimate. If one can only estimate the number of people available for a large construction job, then the time it will take to do the job must also be an estimate.

(g) Algorithm constraints. The number is not in a form that lends itself to computational or other algorithms. For example, it is impossible to write $\pi + 2$ as a single finite decimal, so an approximation must be used.

Here are some further examples, mostly taken from *Applying Arithmetic,* given in a form that might make them appropriate for classroom use. The reader should cover the unread material with a piece of paper because answers are given immediately after the questions.

Examples

1–3. An analysis of data in almanacs shows estimates to be much more common than many people seem to realize. Each of the statements below is taken from *The World Almanac and Book of Facts 1979* and contains an estimate. Which of the seven types of constraints has forced the estimate?

1. Current theories say the first hominid (humanlike primate) was *Ramapithecus,* who emerged 12 million years ago. (p. 719)

Answer: The actual value is unknown (type *a*). It was probably calculated from estimates *(f)*.

2. The area of the Sahara desert is 3,320,000 sq. mi. (p. 447)

Answer: The area varies *(b)* and there is a theoretical limit to the ability to measure it *(c)*.

3. The average July temperature in San Diego, California, is 69°F. (p. 797)

Answer: The answer varies with time *(b)*.

Comment: The mean, median, mode, and other statistics often play roles as estimates.

4–6. Why must the answer to each question be an estimate? Would a single number or an interval be preferred for the estimate?

4. How long does it take to get to the nearest airport from your home?

Answer: The answer varies *(b)*. An interval would be preferred.

5. How much pressure should a car safety belt be able to withstand?

Answer: An overestimate should be given to belt manufacturers for reasons of safety *(e)*. A single number would be preferred.

6. What is the population of the People's Republic of China?

Answer: The population varies with time *(b)*. An interval is more accurate, but a single number (probably the midpoint of the interval) is more convenient.

7. If a school puts an upper limit of 25 students in each English class, how many English classes will be needed for 110 students?

Answer: 5, rounded up to the nearest whole number from 110/25, or 4.4. (This falls under type *d.*)

8. A coin, believed to be unbiased (balanced), is to be tossed 100 times. Give the most likely number of tails that will occur. Give an interval into which the number of tails that will occur will tend to fall about 90% of the time.

Answer: The most likely result is 50, found by multiplying the probability of tails (1/2) by the number of repetitions of tossing (100). Probability tables indicate that for 100 tosses of a fair coin, the number of tails will be from 41 to 59 about 90% of the time.

Comment: The number 50 and the interval 41 to 59 are respectively called the *maximum likelihood estimate* and the *90% confidence interval* for the question: How many tails will occur in 100 tosses of a fair coin?

9. In determining the amount of cement to buy to fill a circular play area 5 yards in diameter and 4 inches deep, what estimates are necessary and why?

Answer: The play area is circular, but the cement fills a cylinder. The volume can be calculated using the well-known formula $V = \pi r^2 h$, where in this case $r = 5/2$ yd and $h = 1/9$ yd. This gives a volume of $25\pi/36$ cubic yards. Estimation is now forced because cement is not sold in such multiples of π (constraint type d). So we must substitute for π either a rational estimate we know, such as 3.14 or 22/7, or a rational estimate stored in a calculator (type g). One scientific calculator gave 2.181661565 cubic yards as the result. This would be rounded up to 2.25 or 2.5 cubic yards, depending on the unit in which cement can be obtained (type e).

Different calculators may yield slightly different results in the computation of Example 9, either because they have stored different approximations to π or because they round differently. Today's computers do not always approximate irrationals by rationals. Symbol manipulator programs such as muMath do not substitute estimates for π unless asked to at the very end of a problem.

REASON 2: ESTIMATES INCREASE CLARITY

> Its [mathematics'] chief attribute is clearness; it has no means for expressing confused ideas.—Jean Baptiste Joseph Fourier
>
> It is a safe rule to apply that, when a mathematical or philosophical author writes with a misty profundity, he is talking nonsense.—Alfred North Whitehead

By *clarity* we mean *ease of understanding*. A school budget of $148 309 563 for a school population of 62 772 pupils might be reported as "about $150 million for 63 000 pupils." A house on a lot whose width was surveyed as 40.13 feet would almost certainly be said to be on a "40 foot" lot. In these cases, the estimate is clearer than the more precise actual figure. Estimates for the purpose of clarity are almost always calculated by rounding.

Clarity and precision are often in conflict; the more precise value is often not as easily comprehended. For example, it is easier to remember that a kilometer is a little over 1600 yards than to recall or memorize that 1 km = 1609.344 yd. Precision is an important idea that should not be underrated, but on many occasions sacrificing precision for clarity is beneficial.

Examples

10–12. The numbers below are taken from actual reports. Indicate how each might be estimated for purposes of clarity.

10. 3 851 809 square miles, the area of Canada.

Answer: 3 850 000 or 3 900 000 or 4 million square miles, depending on the particular circumstances surrounding the need to know this value. The estimate is needed because the short-term memory of most people can retain only a few digits.

11. 4 493 491, the number of scouts and leaders in the Boy Scouts of America (1980)

Answer: 4.5 million. Even exact counts are often estimated for purposes of clarity.

Comment: In actuality, of course, the number given is too precise. The count varied throughout 1980.

12. 8′ 11.1″, the height of Robert Wadlow, reportedly the tallest person who ever lived

Answer: 8′ 11″. Of the examples 10–12, this is the least likely to be estimated because, in records, there is a natural desire to keep as much accuracy as possible. The brief but vague estimate "almost 9 feet" (an interval with ony one endpoint known) has little clarity.

13. On weekdays that are not holidays, national television newscasts always mention the Dow-Jones Industrial Average (DJIA). What does this average estimate?

Answer: The DJIA is based on the worth of the stocks of thirty "industrial" companies in the United States, including General Motors, U.S. Steel, and IBM. It is viewed as an estimate of the economic health of U.S. industry.

Comment: There are other stock averages, some including many more stocks than the DJIA, but because of its long history and the amount of analysis that has been done with it, the DJIA is more familiar, and thus a clearer (even though perhaps less accurate) estimate to many people.

14. On four tests of basic arithmetic facts, Jill scored 8 out of 10, 9 out of 12, 9 of 9, and 7 of 15. Estimate what percent of basic facts Jill knows.

Answer: Adding the numbers of questions and correct responses, Jill scored 33 of 46, or about 72%. (The rounding to the nearest hundredth, i.e., the nearest percent, is for clarity.) If it is felt that the four tests should be given the same weight, then the estimate is the mean of 8/10, 9/12, 9/9, and 7/15. The mean is 181/240, or about 75%.

Comment: Scores on tests are always estimates of knowledge; having a single score for Jill is merely easier to understand or to use (the next reason for estimating) than having four scores.

15. Examine the advertisement reproduced below. Which number is an estimate? What is the actual value being estimated?

CARSONS PRIVATE
LABEL OVERCOAT
SALE
1/3 OFF

Carsons wants to put an overcoat under every Christmas tree! Handsome all-wool coats, hand-picked for fashion, quality and value, in a variety of lengths, looks and colors: both regular and long sizes. Short coats, reg. 180.00, now 118.80. Long coats, reg. 210.00, now 138.60. Long tweeds, reg. 220.00, now 145.20. On sale now thru December 13 or while quantities last, in Men's Outerwear, second floor, Wabash; suburbs except Aurora, Winnetka. Special Christmas Hours: State Street open 'til 7:00; most suburban stores open late, too!

Answer: The number 1/3 is the estimate. The actual discount is 34% for each of the three types of coats mentioned. The reason for the estimate seems to be a view that 34% is not as easy for most people to understand as 1/3. (Note that here we have a case of a fraction being used to approximate a decimal. This is not as rare an occurrence as textbooks would lead us to believe.)

Comments: Carsons could not be accused of misadvertising because they have understated the amount of the discount. Why Carsons would use a 34% discount instead of 1/3 is another question. A possible reason is that under a 34% discount, all even dollar amounts translate to even penny amounts. (This falls under consistency, the fourth reason given here for estimating.) Another possible reason is that it was considered easier to multiply by .34 than to divide by 3. (This reason would fall under facility, the reason given in the next section below.)

16. What single number estimate is often used as a measure of the wealth of a community?

Answer: One common estimate is median family income. It is a better estimate than the mean family income because it is less affected by the wealth of a few very wealthy families. Another common estimate is per capita property value, found by dividing total property value by the population of the community.

Comments: Though a single number estimate is more easily understood than giving lots of data in a complex situation, much information can be lost. No single number gives a picture of the variation of wealth within the community. Either of the common estimates mentioned here misses intangibles (e.g., community pride or physical beauty of the surroundings), services provided from outside the community (e.g., by state or federal governments), and some capital assets (e.g., values of governmental buildings and parks).

REASON 3: ESTIMATES ARE EASIER TO USE

> Mathematicians assume the right to choose, within the limits of logical contradiction, what path they please in reaching their results.—Henry Adams

> Strange as it may seem, the strength of mathematics lies in the avoidance of all unnecessary thoughts, in the utmost economy of thought-operations.—Ernst Mach.

On the one hand, estimating for clarity is in some sense a passive use of estimation; it is not necessary that the estimate be used, only that it be understood. On the other hand, estimating for *facility,* the third major reason for estimating, is an active use of estimation for the specific purpose of simplifying later work or making it more efficient or economical.

This reason for estimating is found in textbooks, but for completeness we still give some examples. When considering how many $3.98 items you can afford to buy, it is easier to do your calculations as if the items cost $4. When a business trip is planned and relative costs of driving versus flying are compared, it is more realistic to use estimates and the calculations are easier: "The trip will be about 800 miles, my automobile gets about 20 miles per gallon, and gasoline costs about $1.20 per gallon; but there will be two days of extra driving, and the motel will cost $35 and meals about $30. Yet the cheapest way to fly costs about $300, and I'll need to rent a car at the destination for five days at $40 a day since I won't have mine . . ."

It is obvious that it is almost always easier in the decimal system to calculate using figures rounded to a particular decimal place than to use exact figures. It is less obvious that even with calculators and computers taking the work out of computation, estimating may make things a lot easier with no important loss in the quality of the answers. In fact, answers derived using suitable estimates may be more reasonable and more realistic than those that attempt to be exact.

Examples

17. The Internal Revenue Service allows (and encourages) taxpayers to round all amounts up or down to the nearest dollar, a practice called "dollar rounding." Give two reasons for this practice.

Answer: The roundings up and down tend to balance out, so there is little difference in totals to be paid. (Half dollars must be rounded up, but can arise in both income and in deductions.) The ease of working with two less significant figures saves the IRS time and money. It is also easier for the taxpayer and leads to fewer mistakes on tax forms.

Comment: A simple exercise is to give students amounts in dollars and cents and ask them to round the amounts as the IRS would allow.

18. Suppose it is decided that federal antipollution funds should be allocated in proportion to the areas of regions of the United States. California

has an area of 158 693 sq. mi. The states of the Northeast have areas in square miles as follows: Maine—30 920; Vermont—9 609; New Hampshire—9 304; Massachusetts—8 257; Rhode Island—1 214; Connecticut—5 009; New York—49 576. Which should receive more funds under this policy, California or all the states of the Northeast?

Answer: California. The easy way to determine this is to round all areas in the Northeast to the nearest thousand. This gives 31, 10, 9, 8, 1, 5, and 50 thousands, the sum of which is 114 000, far less than the area of California. Only if the sum was very close to 158 000 would it then be reasonable to add the more precise areas. But these areas are themselves approximations—in fact, different sources give different values—so being too precise is not appropriate.

Comment: If antipollution funds were allocated in proportion to population, the Northeast would receive more than California.

19. In preparing a $25 000 budget for a year for a family of four, to what nearest amounts should estimated expenses in various categories be rounded? (For purposes of clarity the amount chosen for the budget is divisible by five thousand dollars.)

Answer: $100 seems reasonable. Multiples of $100 are easy to work with, and $100 is only 0.4% of the total.

Comments: Carrying more accuracy than this is seldom worth the effort and leads to frustration when records are not obtainable or when expenses exceed the budgt by small amounts that are of little importance in the total picture. There is a view held by many adults that dealing with estimates in such situations is a weakness and that one should strive for exact amounts at all times. In our opinion, such a view fits the saying, Penny wise, pound foolish.

20. To estimate the product of 15.9 and 3.1 to the nearest whole number, what estimates should you use for the factors?

Answer: None! The product is 49.29, which is closest to 49.

Comment: This simple example illustrates the potential pitfalls of using estimates to approximate or check computations. Despite the given numbers being close to integers and the reasonableness of rounding one number up and the other down, the resulting estimate 16×3, or 48, is not accurate enough.

21. The probability of geting over 60 heads in 100 tosses of a fair coin can be calculated by adding $100!/(61!39!) + 100!/(62!38!) + 100!/(63!37!) + \ldots + 100!/(99!1!) + 100!/(100!0!)$, and then dividing by 2^{100}. But there is a much easier way. What is that way?

Answer: The normal distribution enables an easy approximation. If a coin has probability p of landing on heads and is tossed n times, the mean of the distribution is np and the standard deviation is $\sqrt{(np(1-p))}$. Here $n =$ 100 and $p = 1/2$, so we look at a normal curve with mean 50 and standard deviation 5. Asking what is the probability of getting over 60 heads is akin to asking what percentage of the normal distribution is beyond 2 standard deviations above the mean. The answer is found in tables to be about 0.0228.

Comment: The use of tables for the normal distribution has the semblance of the use of a black box; it seems so far removed from the factorials involved in the precise calculation. But the ease and accuracy of the tables more than compensates.

REASON 4: ESTIMATING GIVES CONSISTENCY

> The critical mathematician . . . believes that there are systems of coherent and consistent propositions, and he regards it his business to discover such systems.—C. J. Keyser
>
> A mathematician, like a painter or a poet, is a maker of patterns.—G. H. Hardy

Government reports on the percentage of people unemployed give that percent to the nearest tenth. For instance, if 8 million of 99 million potential workers were unemployed, the government would report 8.1% unemployed rather than 8.08% or any closer approximation to 8.080808 . . . %. Here an estimate is forced because the original data are inexact, but the particular choice to report in tenths is done to be consistent from month to month and to be consistent with the precision of the raw data from which the unemployment figures are calculated. Similarly, results of calculations from physical measurements should be recorded with significant digits consistent with the input measurements.

Consistency in estimates also arises from conventions or regulations or from a desire to have data of uniform appearance in tables, charts, or graphs. This consistency can lead to some unusual rewriting. For example, a winning "percentage" is customarily a decimal rounded to the nearest thousandth. If a team wins 1 of 4 games, its winning percentage is 25% or .25, invariably written as .250 and read as "two fifty" with the decimal point ignored. This writing and speaking convention is so common that a reader of sports columns who saw either of the equal numbers .25 or 0.250 in a column of winning percentages would probably be confused.

Examples

22–24. Each of the following quantities is normally rounded to the nearest multiple of what place value or power of ten?

22. The inflation rate for a given month

Answer: The nearest tenth of a percent, that is, the nearest thousandth

23. The population of the world

Answer: To the nearest tenth of a billion, that is, the nearest hundred million

24. World records in swimming races

Answer: To the nearest hundredth of a second

In general, when data are reported, the tendency is to have two or three significant figures. However, when records are given, the tendency is to have as many significant figures as can reasonably be measured. In any event, consistency of significant figures is important when estimating, as examples 25–27 show.

25. A rectangular lot is 39.5 meters long and 24.7 meters wide. To achieve consistent accuracy, how should the area of this lot be reported?

Answer: Multiplication of length by width yields 975.65 square meters. Since the given information has only three significant figures, an area of 976 square meters should be reported and the last digit should not be trusted. This can be justified by the following analysis: The given dimensions are themselves estimates. Any value between 39.45 and 39.55 meters would have been reported as 39.5; any value between 24.65 and 24.75 meters would have been reported as 24.7. Thus the calculated area could range between 39.45×24.65 and 39.55×24.75, that is, between 972.4425 and 978.8625 square meters.

26. A gasoline pump and an automobile odometer indicate that 8.5 gallons of gas have been used in traveling 240.2 miles. What is the mpg?

Answer: A calculator gives 28.258824 mpg, which should be rounded to 28 because 8.5 has only two significant figures. Doing the same analysis as in Example 25, one finds that $240.25/8.45 = 28.43$ and $240.15/8.55 = 28.08$, which shows that using a third digit in the answer is not called for by the data.

27. In March 1974, the *Natural History Magazine* gave the following estimates of top speeds (in mph) for animals:

Cheetah	70	Giraffe	32	Pig	11
Elk	45	Elephant	25	Giant tortoise	.17
Rabbit	35	Squirrel	12	Garden snail	.03

What kind of consistency was used in arriving at these estimates?

Answer: With the possible exceptions of the snail and the cheetah, the numbers in the table have two significant figures.

28. In baseball, the earned-run average, or ERA, is a measure of the ability of a pitcher. It is calculated by the formula ERA = $9 \times$ (number of earned runs allowed)/(number of innings pitched). The lower the earned-run average, the better the pitcher's performance. Examine the following news item, referring to the 1981 major league baseball season, which appeared in the *Chicago Tribune,* October 10, 1981:

McCatty's the champ

THE AMERICAN LEAGUE officially named A's pitcher Steve McCatty the earned-run average champion Friday after a review by the official playing rules committee. The rules call for rounding off a pitcher's innings to the nearest complete inning. McCatty had 185 2/3; Sammy Stewart of Baltimore had 112 1/3. McCatty's total was upgraded to 186; Stewart was audited to 112. Had the actual figures been used, Stewart would have been the winner. The final official figures were 2.322 for McCatty and 2.330 for Stewart.

From this information, one can deduce that McCatty allowed 48 earned runs and Stewart allowed 29. What would have been the ERAs had the actual figures been used? Think of two other consistent rounding procedures that could have been used and determine who would have won had they been used.

Answers: Normally in baseball statistics the ERA is rounded to the nearest hundredth, but in all cases here we round to the nearest thousandth. Had the actual figures been used, McCatty's ERA would have been 2.327 and Stewart's 2.323, confirming that Stewart would have been the winner. (Could it be that a rounding procedure is used to avoid division by a fraction?) Another possible procedure is to truncate or round down to the nearest whole number of innings; under this procedure, their ERAs are 2.335 and 2.330, and again Stewart wins. Still another possible procedure is to round up to the nearest whole number of innings; under this procedure, their ERAs are 2.323 and 2.310, and again Stewart wins.

REASONS FOR MAKING A SPECIAL EFFORT TO TEACH ESTIMATION

The previous discussion has elaborated on many good reasons for estimating. However, does estimation merit an important role in the curriculum? Perhaps estimation is so easy and so pervasive that students learn the ideas even without formal instruction. Most leaders today disagree with this view; it is generally felt that the amount of instruction in estimation is not what it should be. (Increasing attention to this subject is one reason for the existence of this yearbook!) But why, if estimation is so widely represented in mathematics, must a special effort be made by teachers to work with these concepts? There are many reasons for such an effort, and these reasons for teaching estimation are not the same as the reasons for doing estimating.

1. *Today's instruction in estimation is often misleading.* Probably the most common type of estimation exercise asks students to estimate the answer to a problem such as 49 × 82 before they do the precise calculation. The idea is to round 49 to 50 and 82 to 80 and then multiply to get the estimate 4000.

However, when this type of problem is first given to students, they usually do not use this idea. Instead, they multiply out 49×82 the long way, getting 4018, and either use 4018 as the estimate or round that answer to get the estimate. The teacher is frustrated. The student has done the problem backward! "You are supposed to estimate first, then calculate! The estimation can help you to check whether the calculation is correct." Most students are unconvinced. Why should they give up a sure-fire way to guarantee a good estimate?

It has often been said that there is logic even behind wrong answers and "wrong" procedures. That is the case in the scenario above, for indeed, the student's procedure is more like the real world than the teacher's advice. Seldom do we estimate when we can easily get an exact answer—with a calculator, for example. Furthermore, as Example 20 above shows, estimating does not always work. So we often perform precise calculations for the purpose of obtaining an estimate. By not dealing realistically with an estimation, today's instruction—either by teacher or through the textbook—often distorts the reasons for estimating, and students do not learn how estimation is used.

2. *Estimation is often the preferred alternative or the only alternative possible in a situation.* It is not by chance that the section of this article dealing with situational constraints is the longest of the four devoted to reasons for estimating. Constraints force estimating a great deal more than the public realizes or textbooks suggest. Whereas people are led to believe that estimating is a weak sister to exact computation, the truth is that estimation is quite often the stronger sister or the only child.

For an example from geometry, we know that an angle cannot be trisected with straightedge and compass. But there are constructions that will yield a close approximation to a trisector. For instance, by constructing five successive bisectors, we can construct an angle whose measure is 11/32 or 21/32 of the measure of any angle, only about 1% off a trisector, and each further bisection can reduce the error by 50%. Things impossible to do exactly may be possible "for all practical purposes." (As a bonus, the mathematics of this particular estimation is quite nice.) By avoiding estimates in situations like these, teachers leave their students with a distorted view of the power of mathematics.

3. *Estimation procedures can lead to new insights about exact procedures.* For example, the polynomial $x^2 - 15x + 51$ cannot be factored over the rationals, but $x^2 - 15x + 50$ can be. Are the solutions to $x^2 - 15x + 50 = 0$ good estimates to the solutions to $x^2 - 15x + 51 = 0$? In general, it is interesting to examine what happens to the solutions to an equation if one of the coefficients is changed just a little.

Let a, b, c, and d be nonzero integers between -10 and 10. What are the largest and smallest solutions possible for $ax + b = cx + d$? How does the answer to this question change if the bounds are $-n$ and n, or if they are 0 and n? These kinds of questions suggest new ways to check solutions.

4. *Wrong procedures may give very good estimates.* We are quick to dismiss procedures that do not give the exact answer. Here is a typical example. Sally can mow a lawn in 40 minutes and John can mow the same lawn in an hour. (Ignore the nonreality of the question in favor of the point that is made for estimation.) How long will it take them to mow the lawn working together? One student argues: "They average 50 minutes, so an estimate of the time it would take them to mow the lawn working together is 25 minutes." This method is wrong, but how close is it to the solution given by the customary method of solving such "work" problems? Procedures "wrong" for exact answers may, in fact, be useful procedures for estimates.

5. *Estimation ideas affect student grading and placement.* This point is not so much a reason for teaching estimation as it is a reason for teachers themselves to be more aware of estimation. Ignorance about estimation can have long-lasting deleterious effects. An individual's score on a test is an estimate of ability in the area of the test. For example, IQ scores are often derived from a normalized scale with a mean of 100 and standard error of measure of 5 points. This means that a student who has an IQ of, say, 100 can be expected to obtain a score between 95 and 105 (100 ± 5) only 68% of the times that he or she completes the test. That the later performance of students often belies predictions based on such tests can be explained to some extent by the large widths of the corresponding intervals. Yet seldom are students' scores on tests treated as estimates. Many teachers will keep a student from a higher grade or from a more advanced class for the want of a tenth of a point. A more complete awareness of estimation could lead to more careful treatment of students.

6. *Ignoring estimation gives students a distorted view of mathematics.* The facet of mathematics that strives for single correct answers covers only a part of mathematics. A mathematics experience devoted entirely to fixed procedures leading to single answers misses the other aspects of mathematics and does a disservice by distorting our subject and its uses. Obsessions with exact answers often force unnecessary calculations and keep people from gaining experience and confidence in estimation judgments. They can kill intuition with detail (the proverbial losing the forest through the trees) and reinforce the false notion that exactness is always to be preferred to estimation.

The appeal of mathematics is in a large way due to the richness and breadth of its concepts, procedures, and applicability. As the preceding discussion attempts to demonstrate, estimation ideas do not weaken this appeal; they enhance it.

The essence of mathematics lies in its freedom.—George Cantor

2

Teaching Computational Estimation: Establishing an Estimation Mind-Set

Paul R. Trafton

SEVERAL articles in this yearbook establish the importance of estimation as a mathematical topic and demonstrate its extensive use in mathematics and daily life. Computational estimation is one of the most powerful and useful aspects of estimation, and building a strong computational estimation strand into school mathematics programs must be a top priority for curriculum developers in the near future. Much work, however, needs to be done if this goal is to be realized. Present programs are extremely limited in their treatment of estimation, and so students do not generally perform well on estimation tasks. Conventional instruction presents highly prescribed ways of working with numbers rounded to multiples of 10, 100, 1000, and so forth, and stresses finding a single "correct" estimate. Not only does such instruction fail to teach students quick, efficient ways of estimating easily, but, even more significantly, it fails to develop (*a*) the sense that estimation is a highly useful tool and (*b*) the sensitivity and flexibility that are crucial to being an effective estimator.

Most students are uncomfortable with estimation. They are not sure what it is or why they need to do it. They find estimating cumbersome and often require paper and pencil and a great deal of time to produce an estimate. When students are required to estimate, they frequently work out the exact answer on paper first and then round it to get an estimate. In fact, many equate estimation with conventional rounding. Further, and more fundamentally, they do not view estimation as legitimate mathematics; in their minds, mathematics deals only with exact answers. They often ask why they can't just find the "real" answer.

To be effective, instructional programs must do two things. First, students must become aware of, and develop proficiency with, several useful procedures for finding estimates. These strategies, which are the focus of article 3 in this yearbook by Barbara Reys, are necessary if students are to learn to estimate quickly, efficiently, and effectively. Second, instruction needs to address a cluster of variables whose purpose is to establish an "estimation mind-set." This includes such things as accepting the legitimacy of estimation, knowing what an estimate is, sensing when it is appropriate to estimate, recognizing how close, or precise, an estimate is required in a given situation, selecting an appropriate strategy, and recognizing whether a computed answer is sensible. Unlike specific strategies, which can be addressed directly in lessons devoted to them, these aspects of estimation are better viewed as threads that need to be woven regularly into instruction in estimation, computation, and problem solving.

Developing students' thinking and reasoning ability in estimation must be given as much attention in programs as techniques for finding estimates. Estimation is a complex skill with many of the same subtleties and complexities as problem solving. Developing an estimation mind-set essentially deals with developing students' reasoning, insight, and decision-making skills with this topic. It also deals with having students accept estimation as a worthwhile skill and developing their confidence in doing it.

Teachers have an important role in teaching estimation. They must carefully motivate the students; pay careful attention to students' thinking processes; make clear the purposes of the work; and be sensitive to the pace of instruction and the level of precision in estimating. Fundamentally, teachers must help students acquire an estimation mind-set. This article addresses four broad aspects of developing this important mind-set and presents suggestions for teaching each of them.

ESTABLISHING THE LEGITIMACY AND USEFULNESS OF ESTIMATION

A major goal for the study of computational estimation is having students come to view it as valid and useful. This viewpoint grows out of an understanding of what estimation is and why it is done as well as knowing how to estimate quickly and easily. Obtaining an answer that is reasonably close to an exact answer seems strange to them at first. Much mathematics instruction necessarily emphasizes finding the exact answer—with no margin for error. It is natural that students may resist the topic at first as a strange and unnecessary intrusion on their orderly world of mathematics.

The following teaching suggestions are important elements in creating the appropriate climate for the study of estimation, establishing its characteristics and usefulness, building confidence in doing it, and making it part of students' thinking. These suggestions need to be included in initial work and emphasized throughout estimation instruction. They are also helpful for teaching specific strategies for finding estimates.

Teaching Suggestions

1. *Introduce estimation with examples where estimated or rounded*

amounts are used. Numbers used in newspaper headlines and articles are frequently approximations rather than exact amounts:

- "Over 35 000 Fans View Final Home Game"
- "Stock Market Soars to 1200 Level"
- "$3.5 Million Spent on Street Repairs"
- "2/3 of Voters Support Higher School Taxes"

This activity helps students realize that approximations are part of daily life and therefore legitimate. This activity can be extended to having students identify whether numbers used in articles appear to be exact amounts or estimates.

2. *Emphasize situations where only an estimate is required.* Estimation is useful both as a check on the reasonableness of exact computation and as an end in itself. Many daily situations require only an estimate. It is sufficient to the football fan to hear that a quarterback who has 18 completions in 35 attempts has completed about 1/2, or just over 1/2, of his passes. A person reading a newspaper advertisement about a sale on suits that are marked down from $118.95 to $89.00 is satisfied to determine mentally that this is about a 25 percent discount. The exercises in figure 2.1 help students recognize situations where an estimate is sufficient.

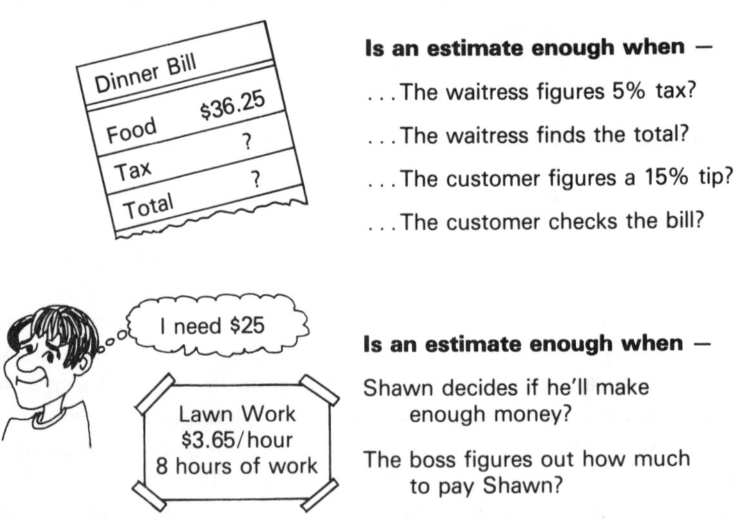

Fig. 2.1

A common textbook exercise asks students first to estimate and then to find the exact answer. Although estimating to check the reasonableness of computation is one important use of estimation, students have difficulty in seeing the purpose of this when first learning to estimate. Since they may lack confidence and proficiency in estimating, they feel that the "real" answer is the important thing—to their teachers as well as themselves. Checking their work by going over it again seems more sensible to them. Thus, students should be given many assignments where they are required only to estimate.

3. *Use real-world applications extensively.* This can help students make estimation a part of their everyday experience. Real-world settings, such as the exercises in figure 2.2, communicate the message that estimation is a useful skill.

Using real-world applications also provides greater encouragement to estimate than the traditional example does. The format in figure 2.3a sug-

PLANT SALE

5″ PLANT	
Coleus	53¢
Swedish Ivy	75¢
Philodendron	99¢
Cactus	$1.43

Estimate. Look for numbers that are easy for you to use.

1. Karen bought 3 Swedish ivy and 1 coleus. About how much did she spend?
2. Phyllis has $10. If she buys 6 cacti, can she also buy 1 Swedish ivy?
3. Dan bought 3 philodendron plants with a $5 bill. About how much change should he get?
4. Pete, Tony, and Carlos together bought 4 plants, one of each kind. They split the cost equally. About how much did each pay?

Estimate the saving on each item advertised:

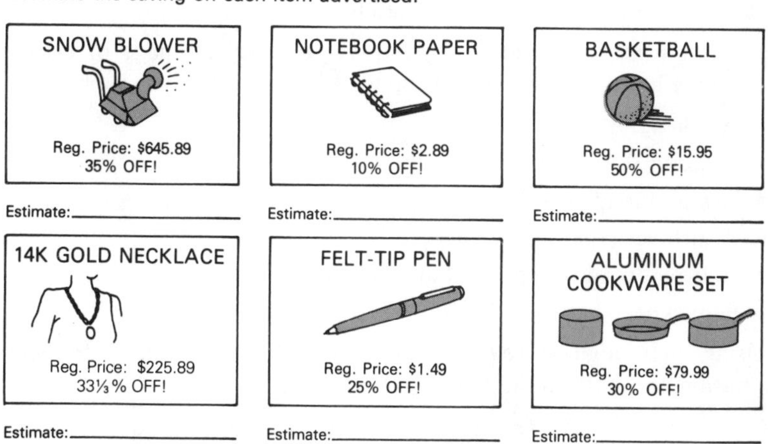

SNOW BLOWER
Reg. Price: $645.89
35% OFF!
Estimate:
NOTEBOOK PAPER
Reg. Price: $2.89
10% OFF!
Estimate:
BASKETBALL
Reg. Price: $15.95
50% OFF!
Estimate:
14K GOLD NECKLACE
Reg. Price: $225.89
33⅓% OFF!
Estimate:
FELT-TIP PEN
Reg. Price: $1.49
25% OFF!
Estimate:
ALUMINUM COOKWARE SET
Reg. Price: $79.99
30% OFF!
Estimate:

Fig. 2.2

gests the use of standard computation, which may actually be easier for students, whereas the same example presented in a real-world setting (fig. 2.3b) can make it easier and more natural to estimate.

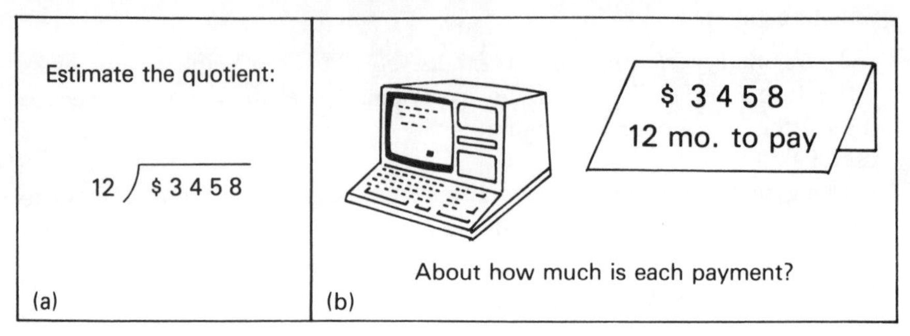

Fig. 2.3

4. *Use easy examples in the early stages, and avoid requiring too much precision in estimates throughout.* Students need to be convinced that estimation is, in fact, easy. The example shown is neither productive nor useful. It requires paper-and-pencil computation and a precise estimate. In daily life it is more common, and just as useful, to think: $15 + $60 + $10. This method can be done mentally and it produces a useful estimate.

Round to the nearest dollar. Then estimate.

```
  $16.28
   57.83
+   8.75
```

Students' abilities vary greatly. Some can estimate with precision by employing unusual strategies and adjusting initial estimates, but many cannot. Thus, pressing for too much precision will make estimation too difficult for some students and they will avoid it.

5. *Emphasize the language of estimation.* The language of estimation needs to be learned. When we estimate, it is common to use such phrases as

- About 12½
- Close to 9
- Just about 15
- A little less than 6.5
- Between 6 and 7, but probably a little closer to 6
- Somewhere between 50 and 60

Using such language regularly does much to communicate the spirit of estimation and help students understand it better.

6. *Accept a variety of answers.* Students must understand that there is no one "correct" estimate. Any estimate that is reasonably close to the exact answer is valid. The emphasis on a single right estimate in present programs

has caused students to view estimation as a formal, rigid process in which the rules must be memorized, just as in conventional computation. This is not what we should be teaching. There are many different ways to estimate, and good estimators have ways of adjusting an initial estimate to get closer to the exact answer. The variety of responses to problems (see fig. 2.4) adds richness to estimation work as students share their estimates and talk about how they obtained them. When multiple responses are encouraged and accepted, we again help students learn more about the process of estimation.

Estimate the total. Be **flexible.** Use a way that is *quick* and *easy to do in your head.*

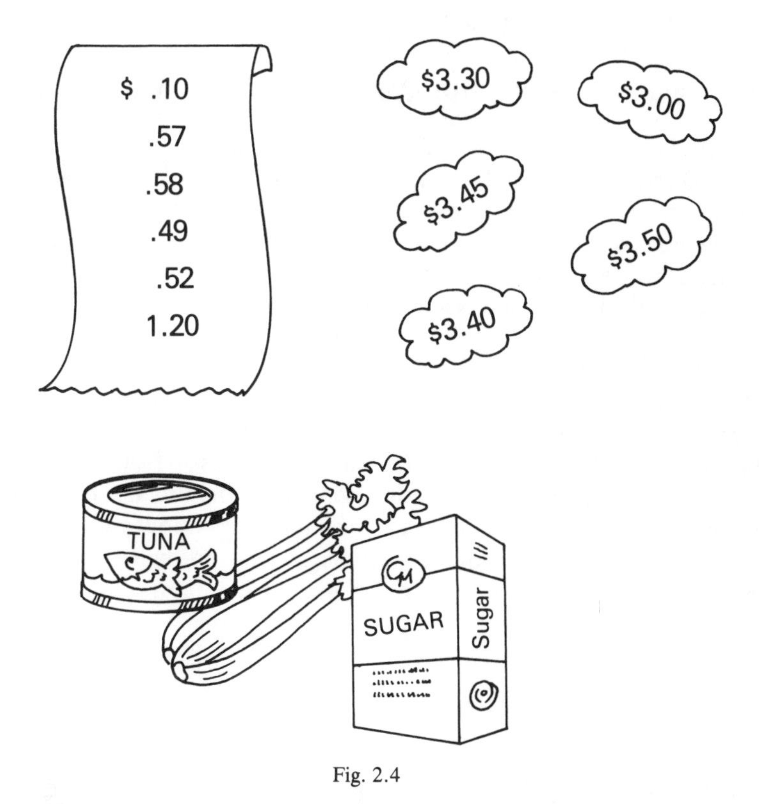

Fig. 2.4

7. *Use oral work and group discussion.* There is a finality about writing answers on paper that can be intimidating. In the early stages of instruction, students will often compute the exact answer on paper and adjust it to a round number. But responding orally encourages students to apply estimation strategies and reduces their anxiety.

Having them share their thinking helps others gain new insights. Oral work also promotes "doing estimation in your head" rather than with paper and pencil. To be truly useful in most daily situations, estimation must be done mentally and reasonably quickly. Also, many students feel that if estimating requires paper and pencil, they might just as well work out the exact answer.

8. *Emphasize estimation regularly.* Estimation is a different way of working with numbers. For students to feel comfortable with estimation and accept it as a normal part of mathematics, it must be done regularly. An isolated lesson at sporadic intervals will have little impact. Several lessons each year should be devoted specifically to estimation, and estimation must be emphasized regularly through exercises on daily assignments and in lessons dealing with computation and problem solving.

DEVELOPING FLEXIBLE THINKING AND DECISION-MAKING ABILITY

Skillful estimation requires flexible thinking and decision making. Decisions must be made about how closely to estimate, whether to overestimate or underestimate, and what strategy to use. Students need practice in analyzing situations with respect to these and other decisions. The goal of such practice is to develop flexibility in thinking and sensitivity to factors that affect how to estimate, such as the context and purpose for estimating.

Teaching Suggestions

1. *Present situations in which students can analyze what type of estimate is needed.* Estimates can range from rough to precise. Sometimes only a ball-park estimate is necessary, but at other times it is important to get closer. In the first situation in figure 2.5, adding the dollars shows that the total is over \$10, and the question can be answered without getting closer. In the second, the sum of the dollars is 8, which is close enough to \$10 to require looking at the cents to see if their sum is over \$2.

Decide:
Eyeball: In a glance can you tell if you have enough?
or
Refine: Do you need to get a little bit closer?

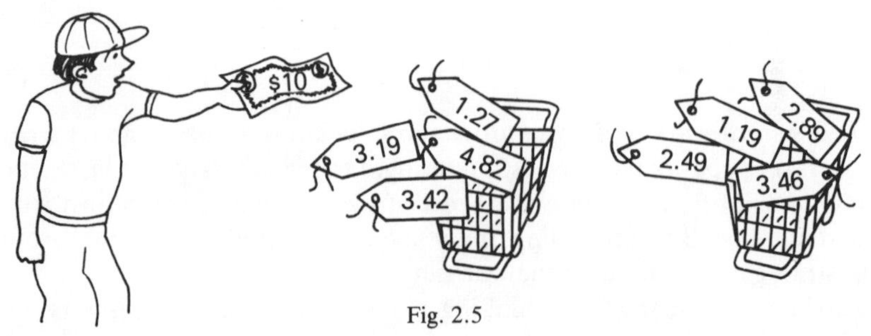

Fig. 2.5

Figure 2.6 shows three types of estimation for a division example. In figure 2.6a the size of the quotient is determined. This is a very useful type of ball-park estimate because it is a check on the reasonableness of the quotient. Finding the first digit of the quotient (fig. 2.6b) adds precision. This method always produces an underestimate, since the remaining nonzero digits in the quotient increase the size of the quotient. A third type (fig. 2.6c) makes use of rounding to the nearest multiple of 6 and produces a closer estimate.

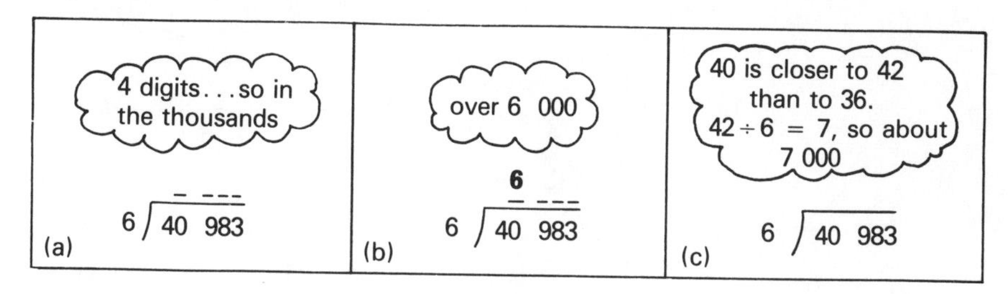

Fig. 2.6

Discussion should emphasize that both ball-park estimates and more precise ones are appropriate, and the choice of which one to use often depends on the situation.

2. *Present examples showing different approaches to the same situation.* This activity promotes flexible thinking, stresses the importance of analyzing a situation to determine an approach that is quick and easy to do mentally, and makes students aware of the diversity that is part of estimation. In figure 2.7 the first student began by adding the dollars and adjusting upward for the cents. The second rounded each number to the nearest dollar; the third rounded each number to the next higher dollar. The latter approach produces an overestimate, which is useful when one has a limited amount of money and doesn't want to risk going over that amount or wants to allow for sales tax.

The two exercises in figure 2.8 make use of fractional parts of a dollar and other shortcuts in addition to more standard ways of estimating. It is valuable for all students to be exposed to these methods, even though only a few may use special techniques.

This work can be followed by presenting "tell how you think" situations, where students describe different methods that might be used for each one.

3. *Present situations where students match strategies to appropriate exercises.* For addition and subtraction there are several useful strategies. Besides learning each one, students also need practice in recognizing when each strategy might be useful or how more than one strategy can be used for the same example (fig. 2.9).

Fig. 2.7

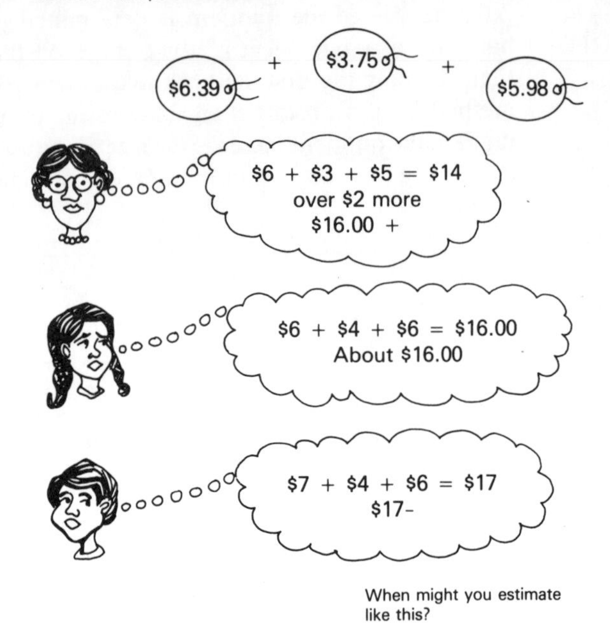

Be a flexible estimator!

Look for short cuts.

Look for ways of **getting closer.**

Look for easy numbers to work with.

How did each person think?

$0.13

About how much will 48 pencils cost?

$\frac{1}{8}$ of 48 = 6.
So, $6.00

100 x $0.13 = $13.00
So, 50 x $0.13 = $6.50

$0.10 x 50 = $5.00

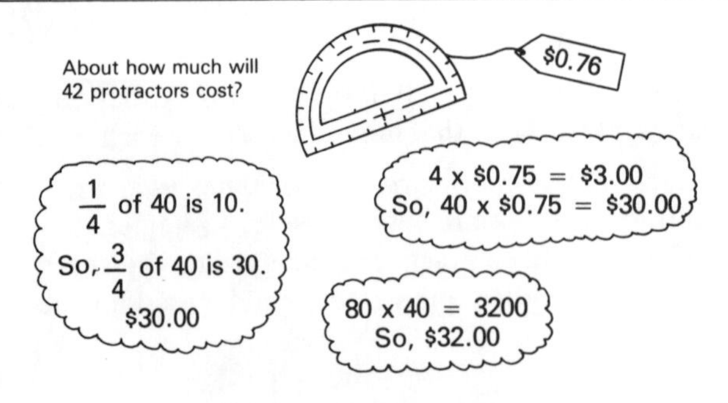

If 40 protractors cost about $30, about how much will 42 cost?

Fig. 2.8

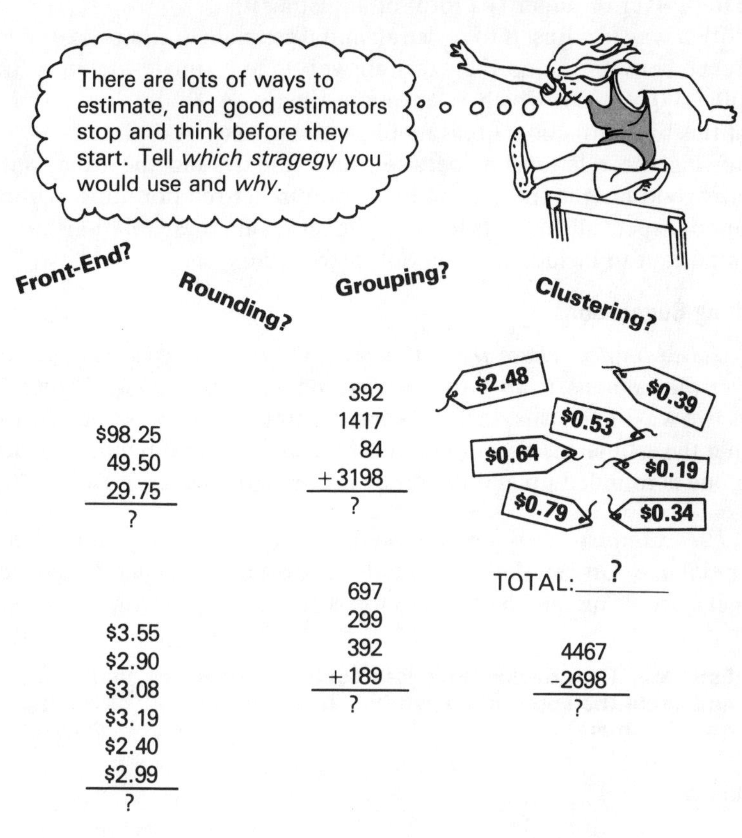

Fig. 2.9

LEARNING TO ADJUST INITIAL ESTIMATES

One important dimension of insightful estimation is having a sense of the relationship between an estimate and the exact answer. It is often useful to know whether an estimate is an overestimate or an underestimate. In the example 38×67, a common estimate is 2800, which is obtained by rounding each factor to the higher ten. Since both factors were rounded up, 2800 is an overestimate. Writing a small minus sign next to the estimate (2800^-) indicates that the exact product is less than 2800. In estimating the amount one saves when purchasing a lamp that originally cost $117.95 and is on sale for ⅓ off, the buyer can estimate by thinking: $120 divided by 3 equals $40. One-third of $117.95 is less than a third of $120.00, so the actual saving is somewhat less than $40. An underestimate, such as 1200 for 32×41, is shown by writing a small plus sign next to 1200 (1200^{+}). Showing whether an estimate is over or under is a form of adjusting it.

Another way to adjust is to revise an initial estimate to compensate for the difference between it and the exact answer. One estimate for 98×325 is 32 500 (100×325), which is an overestimate. A skillful estimator will adjust this by subtracting a reasonable amount, such as 500.

Knowing the relationship between an estimate and the exact answer requires good insight and a good sense of number relationships. Although we cannot expect all students to be skillful at seeing these relationships, it is still important to include them in our instruction.

Teaching Suggestions

1. *Use examples where the estimate is given, and have students show whether the estimate is an underestimate or an overestimate.* Figure 2.10 shows two ways to do this. Include some examples where it is not easy to tell whether the estimate is over or under—for example, multiplying two factors when one is rounded up and the other is rounded down (e.g., 42×29).

2. *Use examples where an initial estimate is adjusted by changing it to a closer estimate.* Give students an initial rough estimate and have them adjust it to get closer. This enables them to focus on just the adjusting process. For

Estimate. Then decide how your estimate should be adjusted, and circle the appropriate symbol. If you can't tell, circle the question mark.

49×28	________	+ ? −	6×328	________	+ ? −
$4102 \div 5$	________	+ ? −	42×38	________	+ ? −

Is the estimate too high, too low, or can't you tell?

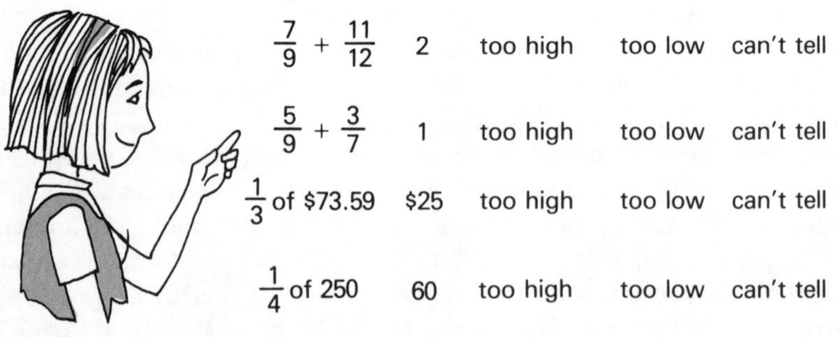

7/9 + 11/12 2 too high too low can't tell
5/9 + 3/7 1 too high too low can't tell
1/3 of $73.59 $25 too high too low can't tell
1/4 of 250 60 too high too low can't tell

Fig. 2.10

the exercises in figure 2.11, first ask how students think the initial estimate was obtained. Then ask them to suggest a closer estimate. Let several students share their revised estimates and tell how they obtained them. Sometimes those who are skilled at mental computation will produce the exact answer this way, and this is also acceptable.

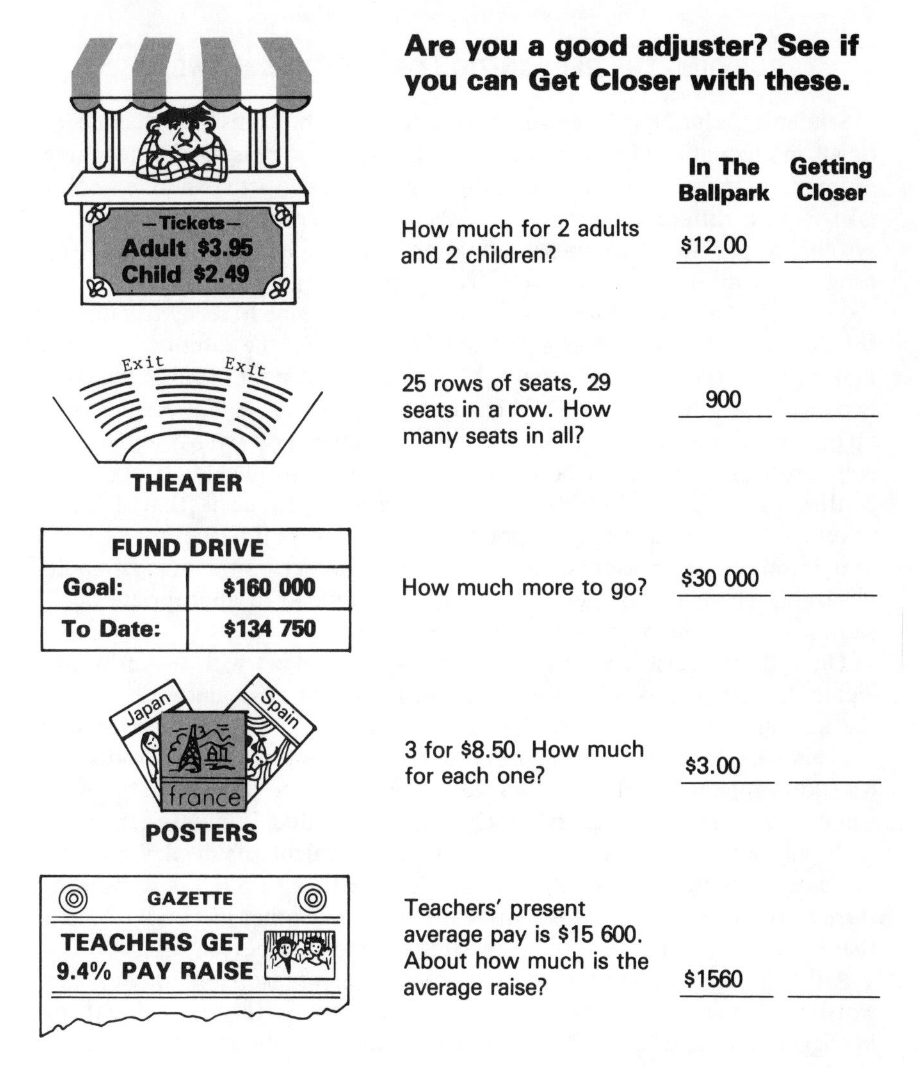

Are you a good adjuster? See if you can Get Closer with these.
In The Ballpark
Getting Closer
—Tickets—
Adult $3.95
Child $2.49
How much for 2 adults and 2 children?
$12.00
Exit
Exit
THEATER
25 rows of seats, 29 seats in a row. How many seats in all?
900
FUND DRIVE
Goal: $160 000
To Date: $134 750
How much more to go?
$30 000
Japan
Spain
france
POSTERS
3 for $8.50. How much for each one?
$3.00
GAZETTE
TEACHERS GET 9.4% PAY RAISE
Teachers' present average pay is $15 600. About how much is the average raise?
$1560

Fig. 2.11

3. *When discussing estimates, ask students about the relationship of their estimates to the actual answer, or have them suggest a closer estimate.* For example, in a problem about a car that traveled 233 miles on 13 gallons of

gasoline, you can suggest an initial estimate of 20 miles a gallon for the mileage and then ask, "Do you think the number of miles per gallon is a little more or a little less than 20? Why?" Make clear that refined estimates are not required and that it is not critical for their estimate to be as close as possible. Again, ball-park estimates are often sufficient when estimating.

BUILDING THE RECOGNITION OF SENSIBLE ANSWERS

Students' failure to recognize when answers are not sensible is a concern of all mathematics teachers and one of the major reasons for teaching estimation. The ability to estimate does enable students to examine a solution from a different perspective. Several teachers who implemented a systematic program of computational estimation have reported that the program made students more sensitive to unreasonable answers.

Yet more than estimation may be involved in learning to recognize unreasonable answers. Sometimes a person just senses that the solution "doesn't look right" without explicitly being aware of estimating. I recall purchasing two blank cassette tapes for $1.29 each and a pen for $0.98. The cashier rang up the items on the cash register and said the total was $4.64. My immediate response was, "That can't be right!" At that point I quickly estimated to confirm by intuitive judgment. This experience suggests that a well-developed sense of number relationships is an important part of the process of determining the reasonableness of a result. In fact, as the next example illustrates, correct estimates will not necessarily lead to appropriate decisions about reasonableness when number sense is absent.

The following task was used in a series of interviews with seventh- and eighth-grade students. They were shown a picture of a calculator displaying the answer 1488 for the example 52×19 and asked if the answer was reasonable. After a significant period of time they estimated the product to be 1000 and concluded that 1488 was close enough to be reasonable. In a situation such as this, more than estimating the product is involved, since in multiplication a difference of 488 between the calculator display and the estimate is not unusual. They need to recognize that 52×19 must be very close to 1000, since one factor is just over 50 and the other just under 20 and that here, a difference of 488 is too large—almost 50 percent of 1000.

Although the factors involved in recognizing unreasonable answers are worth further study, it seems clear that estimation is useful either directly or in triggering thought processes that cause students to monitor their performance when computing.

My experience suggests that students initially need help in distinguishing between checking to see if an answer is unreasonable and checking to see if it is correct. Estimation and related processes are used in the former, whereas the latter involves rechecking the written work. Without careful instruction, students tend to recheck the computation when asked to identify answers that are not reasonable.

Teaching Suggestions

1. *Give noncomputational situations in which students identify or supply a reasonable number.* This type of work, shown in figure 2.12, helps make students aware of what is sensible in various contexts and conditions them to look for reasonable answers when computing.

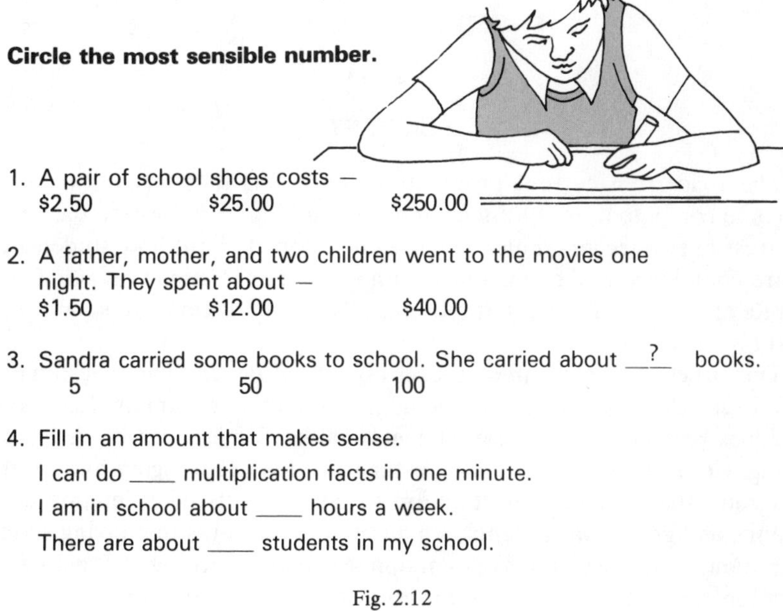

Circle the most sensible number.
1. A pair of school shoes costs –
$2.50 $25.00 $250.00
2. A father, mother, and two children went to the movies one night. They spent about –
$1.50 $12.00 $40.00
3. Sandra carried some books to school. She carried about __?__ books.
5 50 100
4. Fill in an amount that makes sense.
I can do ____ multiplication facts in one minute.
I am in school about ____ hours a week.
There are about ____ students in my school.

Fig. 2.12

2. *Have students identify the number of digits the answer to specific computational examples should contain.* This activity, which focuses on the size of the answer, is a quick and useful check for reasonableness. In the first example (fig. 2.13), the sum must have three or four digits—and more likely three. The third example must have three or four digits, since two two-digit numbers are being multiplied.

3. *Present examples that have been solved, and have students quickly check to find unreasonable answers.* Figure 2.14 shows a quiz that students are to check. Because the answers are displayed, the sole focus is to check for the reasonableness of the answers. A useful variation is giving computational examples with the answers displayed on drawings of calculator windows.

Look at each problem. Don't estimate.
Just decide how many digits there should be in the answer.

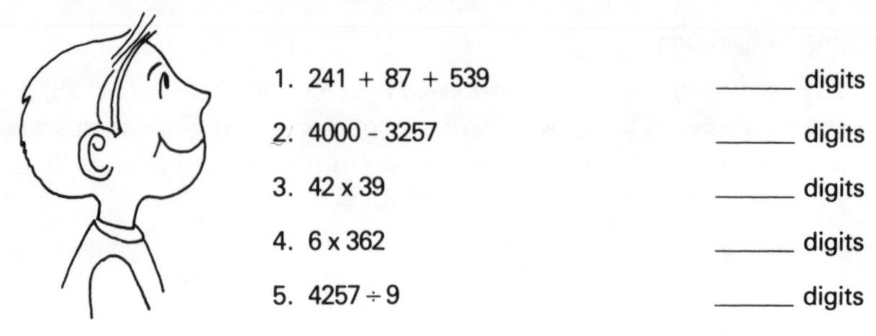

Fig. 2.13

SUMMARY

The road from current programs to programs that teach students to become competent, insightful estimators is a long one. Despite the recent interest of researchers, much needs to be learned about how students acquire confidence and competence with estimation. Estimation is indeed a subtle and complex topic that requires skillful instruction and sound judgment by teachers.

This article has presented several aspects of developing an estimation mind-set, which is one critical element of effective programs. Estimation includes both cognitive and affective variables, which must be developed along with techniques for finding estimates. It is encouraging to note that programs that give careful attention to both aspects of estimation cause change and growth in students. An additional benefit is that giving serious attention in the classroom to estimation makes the teaching of computation more interesting and satisfying for both students and teachers.

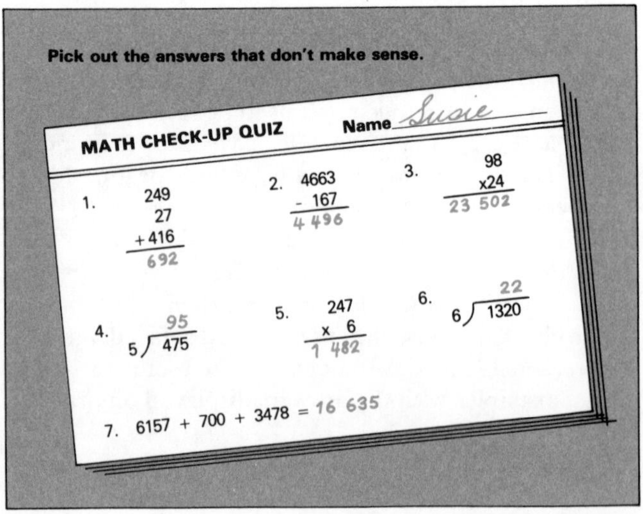

Pick out the answers that don't make sense.

MATH CHECK-UP QUIZ Name Susie

1. 249 + 27 + 416 = 692
2. 4663 - 167 = 4 496
3. 98 x24 = 23 502
4. 475 ÷ 5 = 95
5. 247 x 6 = 1 482
6. 1320 ÷ 6 = 22
7. 6157 + 700 + 3478 = 16 635

Fig. 2.14

3

Teaching Computational Estimation: Concepts and Strategies

Barbara J. Reys

THE very existence of this yearbook as well as the information presented in its various articles documents the support among educators for teaching estimation in elementary and secondary school classrooms. Bridging the gap between knowing that something should be taught and actually teaching it, however, requires concrete answers to questions like these: What specific skills should be taught? How should these skills be sequenced and merged into the existing mathematics curriculum? How should the topic be presented and practiced?

This article will highlight two important features of a comprehensive estimation curriculum: the development of number concepts and the development of estimation strategies. (Many of the ideas and activities illustrated were developed through a National Science Foundation grant entitled "Developing Computational Estimation in the Middle Grades." A complete copy of these materials is available through ERIC [numbers ED 242 526, ED 242 527, and ED 242 528].

THE CONTENT OF AN ESTIMATION CURRICULUM

Estimation, much like problem solving, calls on a variety of skills and is developed and improved over a long period of time; moreover, it involves an attitude as well as a set of skills. Like problem solving, estimation is not a topic that can be isolated within a single unit of instruction. It permeates many areas of our existing curriculum, and to be effectively developed, it must be nurtured and encouraged throughout the study of mathematics. When it is taught as an isolated topic, as we have seen in recent years, the effort may even be counterproductive—leaving students with a general dislike and distrust for the very process. To be truly effective, a careful integration of estimation must occur. A comprehensive estimation curriculum must address several areas:

1. Development of an awareness for, and an appreciation of, estimation
2. Development of number sense
3. Development of number concepts
4. Development of estimation strategies

The first two of these areas are discussed in detail in Trafton's article in this yearbook. The latter two will be discussed and illustrated here.

DEVELOPING NUMBER CONCEPTS

As illustrated in other articles of this yearbook, the results from the National Assessment of Educational Progress (NAEP) estimation items were very disappointing and point out the weak estimation skills of our students as well as a basic conceptual misunderstanding of fractions, decimals, and percents. Unless this conceptual error is corrected, students will continue to struggle with estimating.

Why did 76 percent of 13-year-olds incorrectly estimate the sum of 12/13 and 7/8? Is it because they didn't know how to estimate or because they didn't really understand what 12/13 and 7/8 represent? Very likely it is a combination of the two. How, then, can we help students better develop the notion of fractions, decimals, and percents?

Estimation offers an alternative way of developing concepts related to these numbers. For example, figure 3.1 illustrates a presentation that encourages students to compare numerator and denominator in order to get an estimate for fractions that are less than 1. Emphasis is placed on understanding the size of a fraction relative to 0, 1/2, and 1 before operating with it. This type of presentation can be done before computation with fractions is introduced. The work done in developing estimation with fractions will complement and support later work with computation. Once this basic conceptual development is established, exercises such as estimating 12/13 + 7/8 become simple (see fig. 3.2). A similar format can be used when discussing decimals that are less than 1, as illustrated in figure 3.3. Only after carefully led discussion do students begin to understand that the number of digits contained in a decimal that is less than 1 has little to do with its size. (My students seem to be conditioned to think that "more digits means bigger" !)

Another trouble spot for most junior high school students is percent. This topic is discussed in article 14.

The suggestions made here are not new. The emphasis is on instruction grounded on what the numbers in a problem mean. To me, it is much more important that students understand the meaning of numbers such as 7/8, .48, and 74% than be able to recall the appropriate algorithm to compute with them. In the long run, the algorithms will be forgotten. Numbers like these,

A fraction is close to . . .

- 0 . . . when the numerator is very small compared to the denominator.
- $\frac{1}{2}$ . . . when the numerator is about half the size of the denominator.
- 1 . . . when the numerator is very close in size to the denominator.

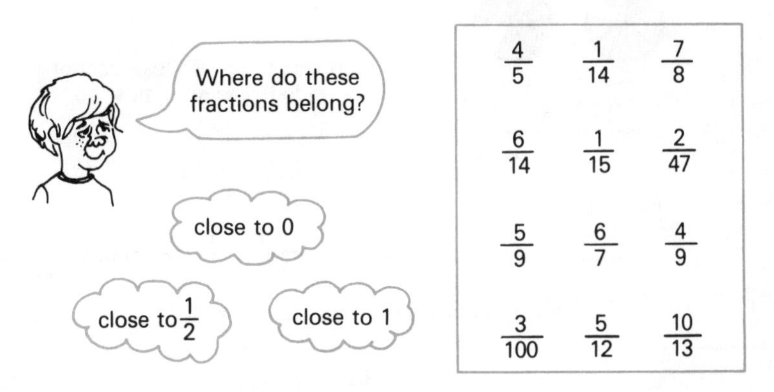

Finish these fractions so that they are close to, but greater than, $\frac{1}{2}$.

$\frac{\ }{8}$	$\frac{\ }{11}$	$\frac{\ }{13}$	$\frac{\ }{21}$	$\frac{9}{\ }$	$\frac{3}{\ }$	$\frac{6}{\ }$

Finish these fractions so that they are close to, but less than, 1.

$\frac{\ }{15}$	$\frac{\ }{9}$	$\frac{\ }{16}$	$\frac{7}{\ }$	$\frac{11}{\ }$	$\frac{9}{\ }$	$\frac{12}{\ }$

Fig. 3.1

however, will be encountered often, and, generally, a good understanding of them together with an ability to estimate will satisfy most needs. It is my observation that time spent developing these basic concepts through a mental computation and estimation approach greatly enhances, and gives meaning to, later work with exact computation.

DEVELOPING ESTIMATION STRATEGIES

Although there are many components to a complete estimation program, one of the most important is the teaching of estimation strategies. The

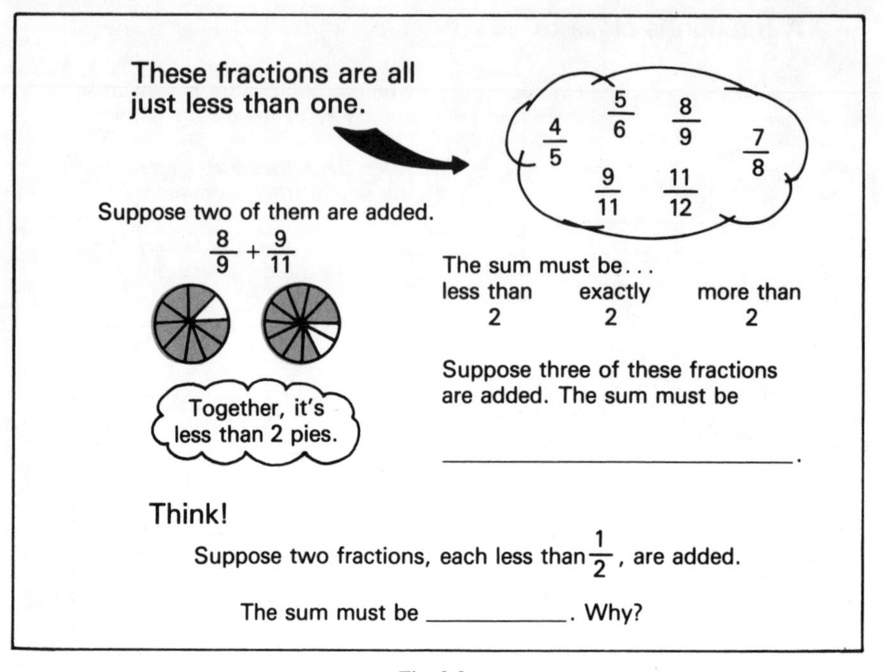

Fig. 3.2

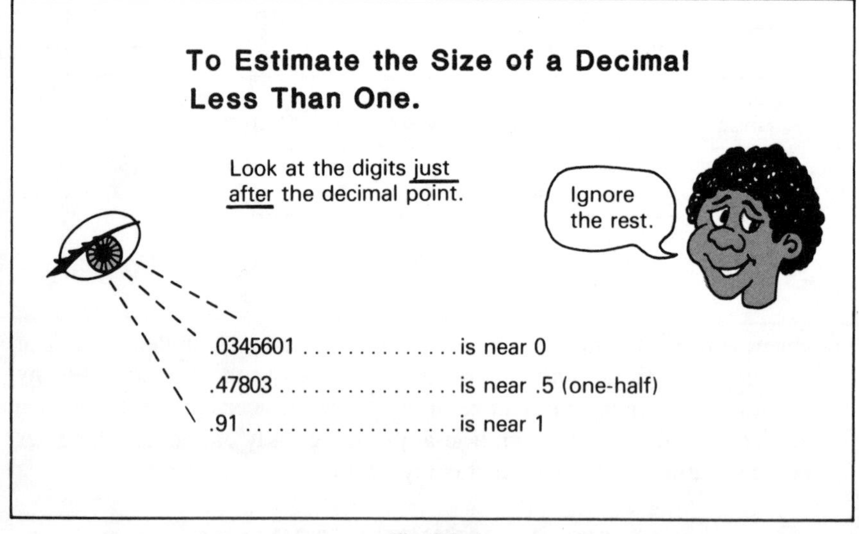

Fig. 3.3

mathematics curriculum of the 1970s and early 1980s included rounding as the core estimation strategy or, more usually, the only strategy. Although this is an important and useful strategy, it is not the most efficient one for many problems. When students and adults who had been identified as good estimators in a recent study were asked to estimate, they used a variety of strategies. They chose these strategies to fit the context of the problem, including the specific numbers and operations involved. Again, this parallels what we know about problem solving. No one problem-solving strategy is efficient for every problem. Part of the task of becoming a good problem solver (or estimator) is being able to select and use a strategy that fits the problem. Recent research has identified several broad strategies that are used by self-developed good estimators. These include the following:

1. Front-end
2. Clustering
3. Rounding
4. Compatible numbers
5. Special numbers

Each of these strategies is best suited to a certain type of problem or operation, and several overlap in their application. Each will be briefly explained and illustrated. A sample format for presenting several of the strategies in a classroom setting will also be included. This format includes a teacher-led discussion of the strategy followed by guided practice.

Front-End Strategy

This strategy is one that even young students can learn to use. Although it can be modified for each of the four main operations, its strongest application is for addition. The focus is on the "front end," or left-most digits, of a number. Because these digits are the most significant, they are the most important for forming an estimate. Front-end strategy is a two-step process. For example, estimate the total of the following:

$1.26
4.79
.99
1.37
2.58

1. FRONT-END... Total the front-end (dollar) amounts.
1 + 4 + 1 + 2 = $8

2. ADJUST... Group the cent amounts to form dollars.

26 and 79 make $1
99 cents make $1
37 and 58 make $1

So, $8 + $3 = $11
initial estimate — adjustment — final estimate

This strategy can be introduced first by using money (as shown here); then other types of numbers can be substituted. The process works well with whole numbers as well as with fractions and decimals:

$$\begin{array}{r} 4316 \\ 1529 \\ +\ 986 \\ \hline \end{array}$$

4 + 1 = 5
or 5000
and about
2000 more . . . 7000

$1\frac{4}{5} + 2\frac{1}{8} + 3\frac{1}{2} + \frac{5}{6}$

1 + 2 + 3 = 6
and about 2 more
or 8

$1.4 + 10.18 + 5.52$

1 + 10 + 5 = 16
and 1 more . . . 17

The adjustment step is a powerful tool for refining an initial estimate; it can be suited to the particular mathematical skill of the user. The real strength of this strategy is in the minimal amount of mental computation involved (only single-digit addition) and the fact that all numbers involved are visible to the user. It works efficiently when applied to a two-addend problem or a six-addend problem. It can be introduced as early as third grade, where the adjustment step may be delayed or treated very casually. For example, to estimate 326 + 214 + 145, a young student could think, "3 + 2 + 1 is 6, so my estimate is a bit more than 600."

As mentioned earlier, the front-end strategy can be applied to subtraction (436 − 285 . . . 4 − 2 is 2, so the estimate is 200, but really less, 200^{-}). However, since with subtraction we are operating with only two numbers at a time, rounding may be at least as efficient.

An illustration of front-end multiplication will likely remind readers that they themselves have used an extended form of this process to mentally compute an exact product.

Front-End	**Adjustment**
4 × 736 . . . 4 × 700 = 2800	4 × 36 makes at least 100
So, 4 × 736 is about 2900.	

The traditional division algorithm is a front-end process. Estimating quotients often produces errors that result in too many or too few digits. Instruction should focus on this trouble spot. For example, estimate the quotient of $7\overline{)3684}$.

Front-End Process

1. Place the first digit of the quotient $\begin{array}{r} 5 \\ 7\overline{)3684} \end{array}$

2. Determine the place value of the quotient $7\overline{)3684}$ with quotient 5--
3. Adjust 500^{+}

The process focuses on the first digit of the quotient and the correct place value of the quotient. This process always produces an underestimate as students quickly see; however, the adjustment step brings greater accuracy to the estimate if it is needed. Figures 3.4 and 3.5 illustrate a classroom format for developing *front-end* for addition and multiplication.

Adjusting Front-End Estimation

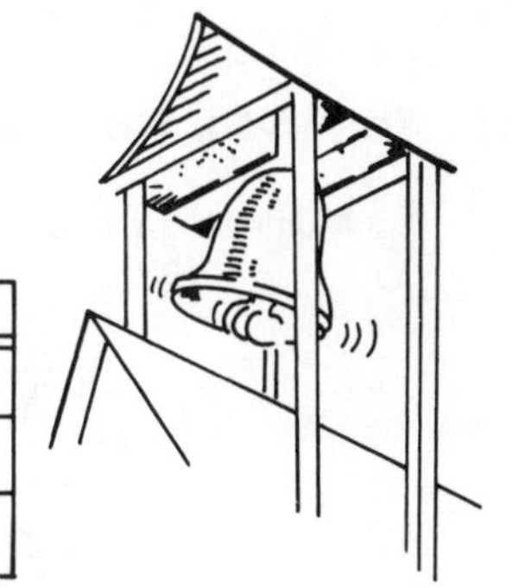

Estimate the Number of Students
Attending the Three Schools.

School	Students
WASHINGTON	378
KING	236
JEFFERSON	442

Remember: With front-end you begin with the digits representing the largest place value!

Add the hundreds: ______________

Estimate the rest: Over 100? Over 200?

Put it together:

CLOSE ENOUGH
9 hundred + 100 +
. . . Over 1000!
GETTING CLOSER
378
236
+ 442
about 150.
I'll say 1050.

To Find an Exact Answer...

Multiplying usually starts at the "back end" of the problem:

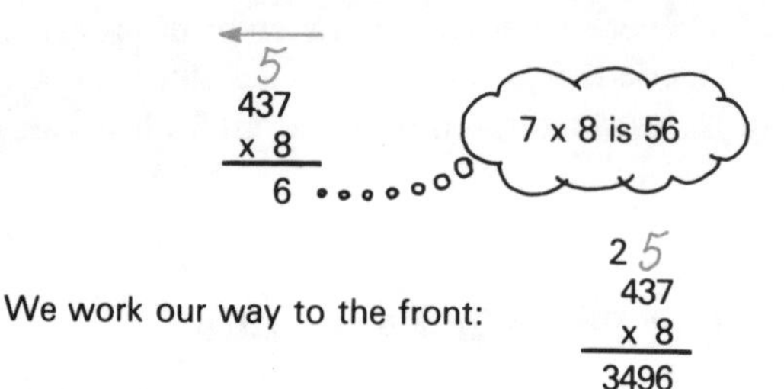

To Find a Good Estimate...

Multiplying starts at the front end:

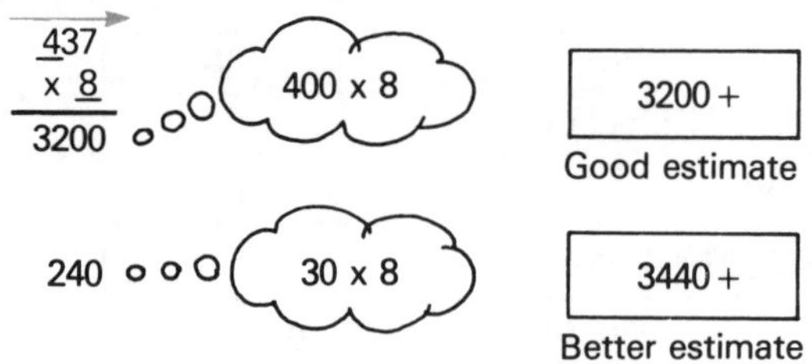

In estimating, use as many digits as will give you the kind of estimate you need.

Fig. 3.5

Clustering Strategy

The clustering strategy is suited for a particular type of problem that we often encounter in everyday experiences. It can be used when a group of numbers cluster around a common value. For example, estimate the total attendance from the following list:

World's Fair Attendance (1–6 July)

Monday	72 250
Tuesday	63 819

Wednesday	67 490
Thursday	73 180
Friday	74 918
Saturday	68 490

Since all the numbers are close to each other in value, we can use clustering to estimate the total attendance.

1. Estimate an "AVERAGE"... {All about 70 000}

2. Multiply the "AVERAGE" by the number of values ... {6 x 70 000 is 420 000}

The strategy can be used with problems involving whole numbers, fractions, or decimals. It eliminates the mental tabulation of a long list of front-end or rounded digits, creating instead a problem with fewer digits that are easily computed. Although clustering is limited to a certain type of problem, it involves a natural translation process and is one that many students (and adults) discover and use on their own.

Rounding Strategy

The rounding strategy is a powerful and efficient process for estimating the product of two multidigit factors. The strategy involves first, rounding numbers; and second, computing with the rounded numbers. A third step of adjustment can sometimes be added when both factors are rounded in the same direction, as illustrated in figure 3.6.

This adjustment of an estimated product is a natural process. I've had a lot of rich discussion with junior high school students about how a product should be adjusted if one factor is rounded up and one down. I always point out that if each number has been rounded in an opposite direction, the process of adjustment has already been accomplished internally and the resulting product is quite satisfactory without further adjustment.

Instruction on this strategy should clearly point out the situations when rounding can be used efficiently. It should also emphasize that numbers can be rounded in many ways, as illustrated in the following examples:

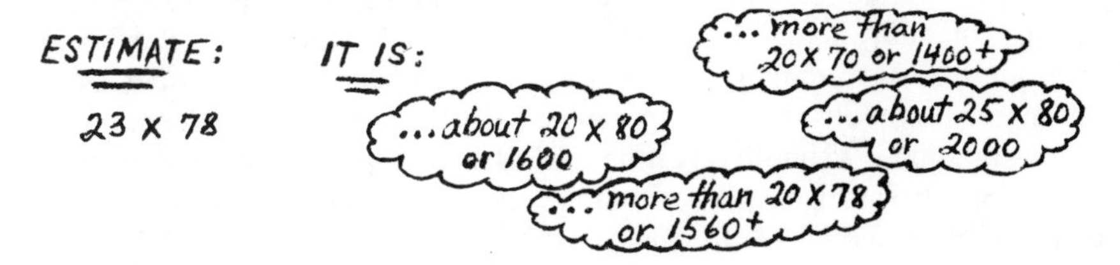

ESTIMATE:
23 x 78
IT IS:
...about 20 x 80 or 1600
...more than 20 x 70 or 1400+
...about 25 x 80 or 2000
...more than 20 x 78 or 1560+

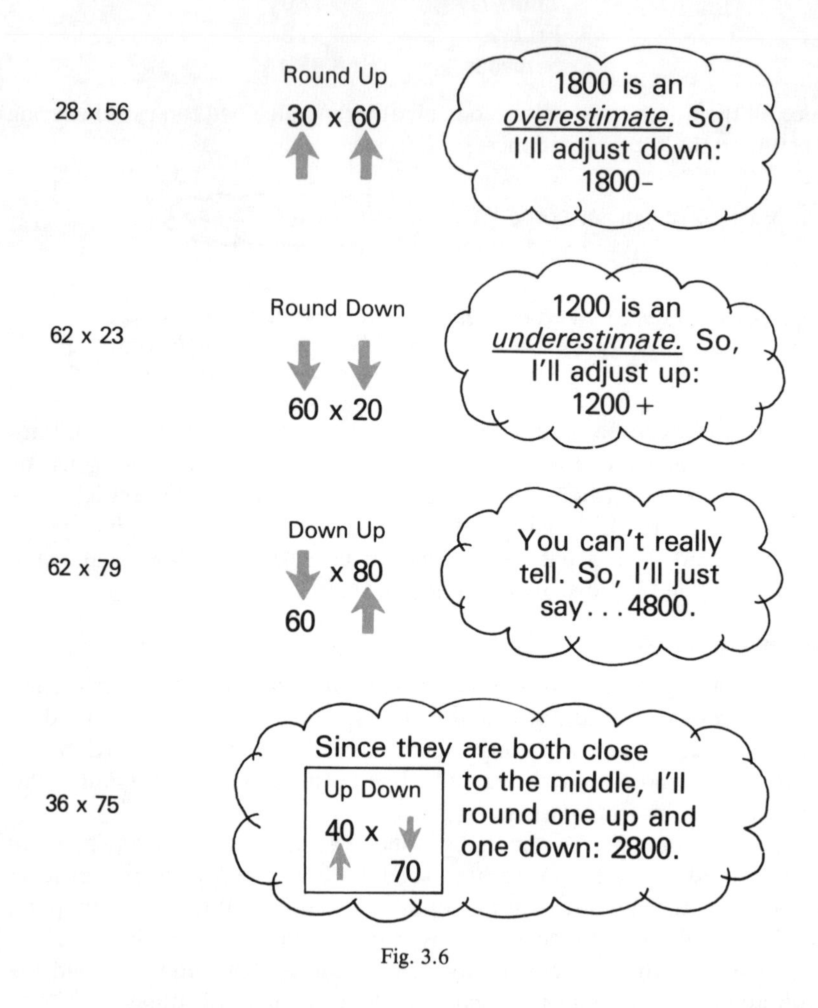

Fig. 3.6

Each rounding choice produces different but reasonable estimates. The choice of rounded factors will be dictated by the user's flexibility and ability to compute mentally.

Students should remember that the purpose of the rounding strategy is to produce mentally manageable numbers. They need to learn to be flexible in their method of rounding—fitting it to the particular situation, operation, and numbers involved. This experience will strengthen the understanding that estimation is a simplifying process. Further, it will help students appreciate that estimators decide for themselves which simpler problem is most comfortable and accurate for their purposes.

Compatible Numbers Strategy

The compatible numbers strategy refers to a set of numbers that can be easily "fit together" (i.e., are easy to manipulate mentally). It encourages the user to take a global look at all numbers involved in a problem and to change or round each number so it can be paired usefully with another number. The choice of the particular set of compatible numbers involves a flexible rounding process. This strategy is particularly effective when estimating division problems. For example:

Estimate:	These **are** compatible sets:	These **are not** compatible sets:
$7\overline{)3388}$	$7\overline{)3500}$	$7\overline{)3000}$
	$8\overline{)3200}$	$7\overline{)3300}$
	$8\overline{)4000}$	$8\overline{)3400}$

The compatible numbers strategy can also be used for addition problems with several addends. The student learns to look for pairs of numbers that "fit together" to make numbers that are easy to compute mentally. Figure 3.7 illustrates this strategy for addition. This strategy involves a certain level of sophistication, experience, and flexibility.

Fig. 3.7

Special Numbers Strategy

This strategy overlaps several already discussed. Students are encouraged to be on the lookout for numbers that are near "special" values that are easy to compute mentally. Special values include powers of ten and common fractions and decimals. For example, each of these problems involves numbers near special values, and therefore they can be easily estimated.

Problem:	Think:	Estimate:
7/8 + 12/13	Each near 1	1 + 1 = 2
23/45 of 720	23/45 near 1/2	1/2 of 720 = 360
9.84% of 816	9.84% near 10%	10% of 816 = 81.6
$.98\overline{)436.2}$	.98 near 1	436 ÷ 1 = 436
103.96 × 14.8	103.96 near 100 14.8 near 15	100 × 15 = 1500

The special numbers strategy is best taught along with the development of fraction, decimal, and percentage concepts, which were discussed earlier in this article. In a sense, special numbers and compatible numbers strategies illustrate best what estimation really is: the process of taking an existing problem and changing it into a new form that has these two characteristics:

1. Approximately equivalent answer
2. Easy to compute mentally

In some instances, numbers are changed only slightly (e.g., 7/8 + 12/13 ≐ 1 + 1). In others, more drastic changes are needed to accomplish characteristic 2 (e.g., 24% of 78 ≐ 25% of 80 = 80 ÷ 4). Several applications of this strategy are seen in figures 3.8, 3.9, and 3.10.

PUTTING IT TOGETHER

Like problem-solving techniques, estimation strategies are developed through careful instruction, discussion, and use. For the best development of estimation skills, the following three phases should be included:

1. *Instruction.* Unless computation estimation strategies are taught, most students will neither learn nor use them. Prerequisite skills (such as the mastery of basic facts and place value) must be reflected in the instruction and development of a strategy. Greater understanding and appreciation of a strategy will result when it is related to different applied situations. Practice is important, but instruction on each of these estimation strategies will complement, direct, and promote meaningful practice.

2. *Practice.* It is important to have a wide variety of practice preceded by specific instruction. Short practice sessions of five to ten minutes each week

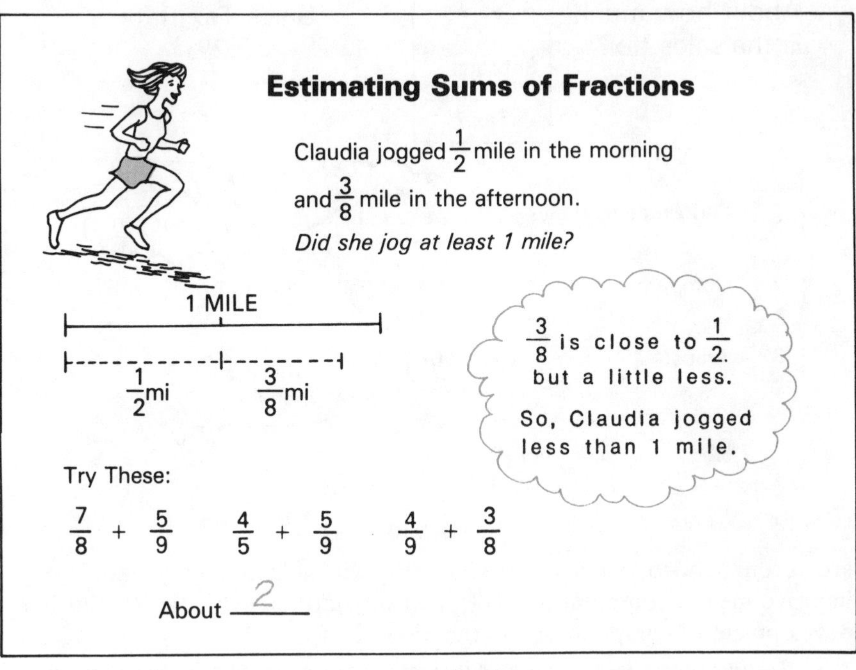
Estimating Sums of Fractions
Claudia jogged 1/2 mile in the morning
and 3/8 mile in the afternoon.
Did she jog at least 1 mile?
1 MILE
1/2 mi
3/8 mi
3/8 is close to 1/2.
but a little less.
So, Claudia jogged
less than 1 mile.
Try These:
7/8 + 5/9
4/5 + 5/9
4/9 + 3/8
About 2

Fig. 3.8

Estimate:
23/49 of 720
Yuk!
Clean the problem up. Think...
• What common fraction is 23/49 near?
It's about 1/2 !
• Change it.
1/2 of 720
• Compute.
That's 360 !
Let's practice. Name a common fraction near each of these:
4/9
5/14
17/18
14/45
7/29
16/30

Fig. 3.9

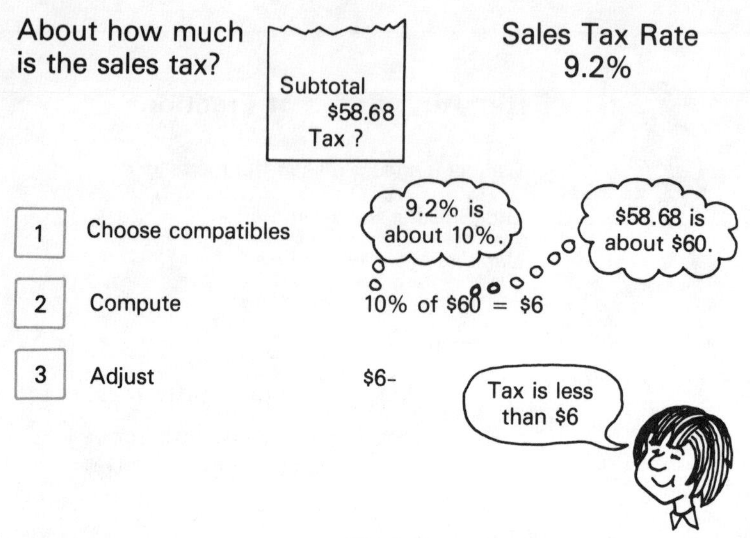

Fig. 3.10

are recommended. Such regular practice will help maintain basic facts, improve mental computation skills, and provide opportunities for further development of computational estimation skills.

3. *Testing.* Periodic testing provides motivation for developing computational estimation. Each test can include about a dozen similar items. An effective format for presentation is to put the items on an overhead transparency and project each problem individually for a short period of time (ten to twenty seconds). Scoring intervals can be set up in advance for each problem, and then selected problems and the strategies used to solve them can be discussed in class.

Estimation is a rich topic and is fun to teach. It offers many avenues of discussion with students which overlap as well as support many other concepts we already teach. The ideas presented here are intended to get you started. Many materials on estimation are becoming available to help guide your instruction. For additional ideas, see the sources listed in the Bibliography.

BIBLIOGRAPHY

Reys, Robert E., and Barbara J. Reys. *Guide to Using Estimation Skills and Strategies, Box I* and *II.* Palo Alto, Calif.: Dale Seymour Publications, 1983.

Reys, Robert E., Marilyn Suydam, and Mary M. Lindquist. *Helping Children Learn.* Englewood Cliffs, N.J.: 1983.

Reys, Robert E., Paul R. Trafton, Barbara J. Reys, and Judy Zawojewski. *Developing Computational Estimation Materials for the Middle Grades,* Final Report. Washington, D.C.: National Science Foundation, 1984. (ERIC Document Reproduction Service No. ED 242 525)

Seymour, Dale. *Estimation, Book A* and *B.* Palo Alto, Calif.: Dale Seymour Publications, 1980.

4

Mental Calculation: Anachronism or Basic Skill?

John A. Hope

IN HIS short story "The Feeling of Power," Isaac Asimov (1962) described a future society where even the most elementary calculations were solved by a computer. People had become so dependent on machines that mental calculation was virtually a forgotten science. A low-grade technician, Aub, startled this futuristic world by managing to reconstruct the ancient methods of calculation. This discovery did not go unnoticed by a select group of politicians and scientists who quickly grasped the significance of Aub's discoveries—the potential liberation of human thought from the intellectual servility posed by the machine.

Shuman, the chief programmer of this advanced technological society, surmised that whoever possessed this resurrected calculative knowledge would have great intellectual and political power:

> Nine times seven, thought Shuman with deep satisfaction, is sixty-three, and I don't need a computer to tell me so. The computer is in my own head.
>
> And it was amazing the feeling of power that gave him. (P. 14)

Unlike Asimov's future society, our contemporary one has not yet progressed to the point where traditional methods of calculation are threatened with extinction. Many studies, including the National Advisory Committee on Mathematical Education (NACOME) (1975) and the British Columbia mathematics assessments (Robitaille and Sherrill 1977; Robitaille 1981), have reported that schools still spend considerable time teaching children how to compute. Nevertheless, the nearly universal availability of hand-held calculators seems destined to challenge the favored position enjoyed by the traditional methods of paper-and-pencil computation currently taught in school.

The calculator far surpasses any previous invention in reducing the memory requirements of the user. To calculate a product such as 123 × 456, for example, requires the recall of neither numerical equivalents nor calculative method. Other than knowing the order of the keystrokes and recalling which entries have been made, the user has been freed from the more taxing memory demands normally associated with calculation.

The calculator holds an obvious advantage over more conventional methods of calculation, including paper-and-pencil and mental. Thus, this question becomes important: Has mental calculation as an educational goal become obsolete, or are there benefits to be gained by systematically introducing a child to mental calculation? The purpose of this paper is to consider that question.

MENTAL CALCULATION AS A PRACTICAL LIFE SKILL

Curricula can be evaluated in any number of ways. One popular method has been to determine the social utility of a curriculum's content. According to this viewpoint, a good curriculum shoud be based on those skills needed to solve problems encountered through everyday activity. Commonly, those skills have been identified through the study of people as they proceed through their regular activities. For example, Wandt and Brown (1957) attempted to identify the routine uses of calculation by asking people to note their off-the-job calculative activities over a twenty-four-hour period.

Each reported use of a calculation was placed in one of four categories: mental-exact, mental-approximate, paper-and-pencil-exact, and paper-and-pencil-approximate. The study reported that approximately 75 percent of the routine calculations were performed using mental procedures, whereas paper-and-pencil methods were used for only 25 percent. Surprisingly, mental-exact calculations were found to be much more numerous than mental-approximate methods. Wandt and Brown concluded that considerable emphasis should be placed on mental-exact and mental-approximate mathematics at both the elementary and secondary levels. This emphasis on mental calculation should, the authors stated, "include not only practice in the mental arithmetic process, but also applications to life situations" (p. 153).

Certainly, curriculum developers must exercise caution when studies of routine behavior are used to judge the merits of a curriculum. Basing curricular decisions on studies of contemporary life can lead to a stagnant and rapidly outmoded curriculum. The identification of a set of mathematical skills whose utility remains timeless is a particularly difficult enterprise owing to the increasingly dynamic nature of society.

The ubiquitous hand-held calculator would undoubtedly affect the findings of a replication of Wandt and Brown's 1957 study. Likely, today's citizen would not use mental calculation methods as frequently as the subjects of Wandt and Brown did twenty-five years ago. Nevertheless, mental calculation can have great practical utility despite the continual proliferation of calculators. It will remain a convenient tool to determine an exact solution for a narrow range of numerical problems, and it is a crucial component of estimation which, as other articles in this yearbook clearly point out, is a valuable skill to develop.

Mental Calculation and Exact Solutions

Teachers have always spent a great deal of time teaching children how to calculate. Major surveys, including Priorities in School Mathematics (NCTM 1981), Overview and Analysis of School Mathematics (NACOME 1975), and the British Columbia mathematics assessments (Robitaille and Sherrill 1977; Robitaille 1981), have concluded that most teachers spend the majority of arithmetic instructional time in pursuit of this goal.

In schools, computational facility is understood to mean proficiency with the conventional paper-and-pencil algorithms. These methods of ciphering have been taught because they are believed to be life skills that are essential for functioning in the adult world. Yet these skills taught in school are not necessarily the skills that people use outside school.

In the study described above, Wandt and Brown (1957) found that the great majority of everyday calculation problems were solved using mental techniques rather than the paper-and-pencil ones emphasized in school arithmetic programs. More recently, Maier (1977) claimed that adults solve most everyday calculative tasks by applying methods quite different from those taught in school mathematics classes. Referring to these unconventional and often untaught procedures as "folk math," he wrote:

> Some of the general differences between school math and folk math are clear. One is that school math is largely paper-and-pencil mathematics. Folk mathematicians rely more on mental computations and estimations and on algorithms that lend themselves to mental use. When computations become too difficult or complicated to perform mentally, more and more folk mathematicians are turning to calculators and computers. In folk math, paper and pencil are a last resort. Yet they are the mainstay of school math. (P. 86)

Are Maier's observations correct? If they are, the modern calculator and the expected advances in calculative technology will pose more of a threat to paper-and-pencil than to mental methods of calculation. If a calculation task seems complex, the user is more likely to reach for a calculator than paper and pencil. But if the calculation task is reasonably simple, what can be more convenient than mental calculation? As Maier has suggested, "Other computation tools may not always be available, but folk mathematicians always carry their brains with them" (p. 89).

Unfortunately, when a mental calculation is required, many children and adults do not seem to carry their brains with them. Some of the young adults sampled by the 1983 National Assessment of Educational Progress even had difficulty with straightforward mental calculations. When asked to multiply numbers like 90 and 70 "in the head," only 55 percent of the seventeen-year-olds were able to do so (p. 32). Fifty-five percent were unable to calculate an exercise like 4×625 mentally (p. 32). On orally presented mental division exercises, almost 40 percent were unable to complete the solutions to items similar to 480/16 and 3500/35 within a ten-second time period (p. 11).

Many educators have realized that mental calculation is an integral part of mathematics learning, far too important to be left to the vagaries of incidental mathematical experience, and attempts are being made to redress the inattention this mode of calculation has received over the years. After identifying the mathematics required in higher education, employment, and general adult life, the authors of the British report *Mathematics Counts* (Cockcroft 1982) suggested that mental calculation should be given a far more prominent position in mathematics programs than it has been accorded recently. This report concluded that the "decline of mental and oral work within mathematics classrooms represents a failure to recognize the central place which working 'done in the head' occupies throughout mathematics" (p. 75).

Mental calculation is an admittedly limited tool for solving most types of calculative tasks, especially when compared to the capabilities of the calculator. Few individuals will attempt to solve a problem such as 123×456, for example, by using mental procedures when a calculator is available. Nevertheless, to expect most people to use some form of mental calculation to solve such problems as 9×300, \$1.99 + \$5.99, 480/16, $1000 - 501$, 2×555, and even 25×48 does not appear to be an unrealistic goal. The challenge for teachers will be to help children develop the necessary capabilities.

Estimation as a Form of Mental Calculation

Many writers, including some in this yearbook, have argued recently that estimation as a method of determining the reasonableness of a proposed solution to a calculation problem should become an important goal of arithmetic programs because of the increased use of calculators. The need for teachers to emphasize mental estimation was discussed in NCTM's *An Agenda for Action* (1980):

> Teachers should incorporate estimation activities into all areas of the program on a regular and sustaining basis, in particular encouraging the use of estimating skills to pose and select alternatives and to assess what a reasonable answer may be. (Pp. 7–8)

Similar recommendations have been made by Trafton (1978), Levin (1981), and Reys (1984).

Despite the claimed importance of estimation, children and adults are not good estimators. For example, only about one-half of seventeen-year-olds could estimate a sum to the nearest million dollars when given four hypothetical government expenditures (Carpenter et al. 1978, 30). The 1983 National Assessment of Educational Progress reported that, at all ages, the performance involving an estimation of a computation was considerably poorer than corresponding computation exercises (p. 31).

Why is providing a quantitative estimate so difficult for children and adults? Some authors (Reys 1984) have concluded that estimation simply hasn't been taught in schools. If this conclusion is correct, the disappointing findings reported by many studies merely reflect a lack of instructional opportunity.

Other studies have looked more closely at the process of estimation, examining the strategies that estimators use. One study (Reys et al. 1982, 197–98) reported that good estimators had the following common traits among others:

> . . . the *quick and efficient use of mental computation* to produce accurate numerical information with which to formulate estimates. All estimators exhibited well-developed skill with multiples of 10 or a limited number of digits, and many others were fluent in mentally computing with larger numbers, more digits, and even different types of numbers (e.g., fractions). On some problems, subjects resorted to *mental computation* rather than using an estimation technique. For these problems and these students, it was more efficient for the person to compute mentally rather than estimate. (italics added)

The close relationship between estimation and mental calculation established by Reys and his colleagues implies that no instructional program purporting to teach estimation skills can afford to ignore the teaching of mental calculation. Estimation in computation, after all, is no more than a form of "less precise" mental calculation.

MENTAL CALCULATION: A WAY OF THINKING ABOUT NUMBER

Providing for the mathematical development of children should involve much more than simply ensuring that each child leaves school equipped with a set of skills and procedures designed to solve everyday problems. The study of arithmetic should help children develop some measure of quantitative thinking, namely, a way of thinking about, and reasoning with, numbers. In particular, children should develop a view that arithmetic is rich in meaning.

As early as 1935, Brownell urged that meaning or seeing sense in what is being learned should be the central focus of arithmetic instruction. He explained that

> the "meaning" theory conceives of arithmetic as a closely knit system of understandable ideas, principles and processes. According to this theory, the test of learning is not mere mechanical facility in "figuring." The true test is an intelligent grasp upon number relations and the ability to deal with arithmetical

situations with proper comprehension of their mathematical as well as their practical significance. (P. 19)

Even after five decades Brownell's fond hopes have not been realized, for many children and adults still do not view arithmetic as a particularly meaningful subject. In a recent analysis of the results of the first two National Assessments of Educational Progress (Carpenter et al. 1981), the authors concluded that although most students possessed an adequate mastery of the mechanical aspects of computation, "many of the skills appear to have been learned at a rote, superficial level" (p. 25). The 1981 British Columbia mathematics assessment (Robitaille 1981) reported similar conclusions about the apparent inability of students to reason with numbers: "Although students perform satisfactorily on computational skill items, results are weaker in areas involving what might be termed 'numeracy' " (p. 2). Computation is seen by most children and adults as a way of "getting a correct answer"; whether the answer makes sense or not is of little concern to the majority of users.

Certainly there has been an attempt by textbook authors and teachers to help children understand how number principles and relationships are involved in the working of an algorithm. For example, most text series include a number of lessons devoted to teaching place-value concepts, the distributive law, and the annexation algorithm (multiplying by a multiple of a power of 10) well before the conventional form of the multiplication algorithm is introduced.

Unfortunately, no matter how well the understanding of these algorithms is promoted, the very nature of paper-and-pencil algorithms likely contributes to a fragmentary view about numbers and number relationships. Efficient ciphering requires that the user parse the original operands into a series of more tractable subproblems. Solving a problem such as 25×48, for example, involves reasoning that can be described in the following way: Eight times five is forty. Record the zero and carry the four. Eight times two is sixteen; adding the carried four makes twenty. Place the twenty to the left of the zero. Four times five is twenty. Place the zero under the first group of figures and one place over to the left; carry the two. Four times two is eight; adding the carried two makes ten. Place the ten to the left of the zero. The answer, after adding, is 1200. This example demonstrates that the proficient user works not with numbers and relationships but with digits and bookkeeping rules. Number relationships must be discarded as so much "intellectual flotsam" because they hinder efficiency.

This pursuit of ciphering efficiency is costly, however, because an overreliance on paper-and-pencil methods can lead to a type of inflexible behavior that could be described as "calculative monomania"—the tendency to ignore number relationships useful for calculation and, instead, resort to more cumbersome and inappropriate techniques. The following calculations are personal observations of this rule-bound calculative behavior:

```
 0           5          0 9 9  9
 1 14       3 6 10     $1 10 10. 10 10          1 2 5
 - 6       -3 5 9     -   9 9. 9 5            x 1 0 0 0
 ----      ------     -----------             ---------
 0 8        0 0 1      0 0 0. 0 5                 0 0 0
                                                0 0 0
                                              0 0 0
                                            1 2 5
                                            -----------
                                            1 2 5 0 0 0
```

Although each of these tasks could have been solved by an elementary mental calculation, the conventional paper-and-pencil algorithm was employed despite its apparent unsuitability. When asked why she went through a lengthy approach to do a calculation that could have been done more easily "in the head," one girl replied defensively, "That's what we have to do in school!" Such rigid behavior suggests that children are enslaved to a technique and that either alternative methods of solution are not known or they have been rejected by the user as being somehow "improper."

Maier's contention that many young adults "are enslaved to the slow and awkward procedures learned in school" (1977, 92) seems to be an accurate description of the unskilled mental calculators who participated in a recent study (Hope 1984). They relied almost exclusively on a paper-and-pencil mental analogue to solve calculation problems involving the products of multidigit factors. Dependence on this method was alarming at times. For example, one subject tried to solve the problem 50 × 64 by "mentally placing 50 on the 'bottom' because I knew it would be all zeros on the top." Further questioning revealed that she imagined the partial products aligned in the manner indicated below:

```
   64
 × 50
 ----
   00
 320
 ----
```

An overdependence on paper-and-pencil methods obscures many number properties that can be used to aid a calculation. For example, the unskilled mental calculators did not think of 8 × 99 as 800 − 8 or 50 × 64 as 1/2 of 6400. These calculations were viewed instead as tasks requiring a routine series of right-to-left, digit-by-digit mental manipulations.

Even obvious number relationships were not used by some young adults to simplify a calculation. For example, one unskilled subject had difficulty with the seemingly simple problem of 8 × 25. She explained, "8 × 5 is 40, and carry the 4; 8 × 2 is 16, . . . 17, 18, 19, 20, 21, 210." Although further questioning revealed that this subject knew 4 × 25 equaled 100, she was unable to use this knowledge to solve the related problem 8 × 25. Not surprisingly, very few subjects who used the awkward paper-and-pencil mental analogue were able to solve more complex problems such as 25 × 120 and 25 × 480.

The use of the right-to-left, digit-by-digit, paper-and-pencil mental analogue was often accompanied by conspicuous physical motions that indicated the subject was attempting to "write" each stage of the calculation. Fingers were used to form calculations either in the air or on a desk. After attempting to solve 12 × 15, one subject explained: "I wrote it down, and then I forgot it. I did 120; and then as soon as I did 60, I forgot 120. The second time I put an emphasis on the 60 by pressing my finger down harder." The reluctance to discard even the physical actions associated with written methods reflects how the thinking of the unskilled subjects was dominated by this particular calculative procedure.

The proposed solutions of the unskilled subjects frequently bordered on the absurd. For example, one subject offered 309 as the solution to 24 × 24. In a second attempt she obtained 5496. When asked which answer seemed to be the most reasonable, she commented: "Probably the second one (5496). I went through it again; and if you're multiplying digits that big times each other, you should probably get a 4-digit and not a 3-digit number."

The difficulties of the unskilled subjects were not due to the lack of a computational tool. Rather, these subjects seemed unaware that different tasks called for different tools. Paper-and-pencil techniques were used to solve mental calculation tasks in the same crude manner as an inexpert handyman might use a wrench instead of a hammer to drive a finishing nail. Though both tools can do the job, one is more awkward, ungainly, and hazardous.

How different the skilled mental calculators seemed. Skilled mental calculation demands that the user "search for meaning" by scanning the problem for salient number properties and relationships. A skilled mental calculator views calculation in the same manner as an artist views a painting. Both individuals see relationships unnoticed by the "untrained eye." To carry this comparison further, color, form, and space cue the artist; number properties cue the proficient mental calculator. The reasoning offered by one young skilled subject demonstrates how the choice of a strategy can be influenced by the apprehended properties.

8 × 99	I did 800 − 8 = 792.
25 × 480	Well, 25 is 100 divided by 4. So I divided 480 by 4 to get 120 and multiplied that by 100.
16 × 16	It's a fact. I know many powers of 2 from working with computers and the binary system.

32×32	Well, 32 is 2^5, and this squared is 2^{10}; therefore, 1024. I just know 2^{10} is 1024.
24×24	24×20 is 480, and 4×24 is 96. 480 and 96 is $580 - 4$; so the answer is 576.
23×27	I did 25^2, which is 625, and subtracted 2^2, or 4.
15×48	10×48 is 480; 1/2 of 480 is 240, and $480 + 240$ is $480 + 200$. And 680 and another 40 is 720.

CONCLUDING REMARKS

Mathematics has been said to contain much that will neither hurt one if one does not know it nor help one if one does know it. Some people may argue that this epigram is an apt description of mental calculation as well. The trend to create devices designed to replace most of one's calculative efforts will no doubt continue, and there will likely be little incentive to acquire considerable expertise in mental calculation. Why should calculative strategies be learned and numerical equivalents be memorized when this information can be purchased for a few dollars?

Despite the obvious advantages that hand-held calculators hold over all other forms of calculative methods, many situations exist in which either an exact or an approximate mental calculation is the preferred mode of calculation (for example, see the article by Usiskin in this yearbook). Moreover, proficient mental calculation involves a type of thinking that will not be replaced easily by advances in technology. Proficiency demands that the user search for meaning and understanding, and any process that depends on the integration of seemingly disconnected mathematical concepts and rules is well worth retaining as a goal of mathematics education. The benefits of mental calculation, therefore, go beyond the ability to make quick and accurate calculations. Max Beberman made this point over twenty-five years ago:

> Mental arithmetic . . . is one of the best ways of helping children become independent of techniques which are usually learned by strict memorization. . . . Moreover, mental arithmetic encourages children to discover computational short cuts and thus to gain deeper insight into the number system. (Sister Josephina 1960, p. 199)

To close on a somewhat museful note, proficient mental calculation involves a sophisticated form of mathematical thinking that must not be allowed to atrophy to the level evident in Asimov's future society. A correct answer may be the destination of a mental calculation, but it is the path followed rather than the destination that reveals a traveler's understanding. A good mental calculator is able to travel many more paths than the poor mental calculator. The ideas contained in this yearbook should help teachers open more paths to more children.

REFERENCES

Asimov, Isaac. "The Feeling of Power." In *The Mathematical Magpie,* edited by Clifton Fadiman, pp. 3–14. New York: Simon & Schuster, 1962.

Brownell, William A. "Psychological Considerations in the Learning and the Teaching of Arithmetic." In *The Teaching of Arithmetic,* Tenth Yearbook of the National Council of Teachers of Mathematics, edited by W. D. Reeve, pp. 1–31. New York: Bureau of Publications, Teachers College, Columbia University, 1935.

Carpenter, Thomas, Terrence G. Coburn, Robert E. Reys, and James W. Wilson. *Results from the First Mathematics Assessment of the National Assessment of Educational Progress.* Reston, Va.: The National Council of Teachers of Mathematics, 1978.

Carpenter, Thomas P., Mary K. Corbitt, Henry S. Kepner, Mary M. Lindquist, and Robert E. Reys. "National Assessment." In *Mathematics Education Research: Implications for the 1980s,* edited by Elizabeth Fennema, pp. 22–40. Alexandria, Va.: Association for Supervision and Curriculum Development, 1981.

Cockcroft, Wilfred H. *Mathematics Counts.* London: Her Majesty's Stationery Office, 1982.

Hope, Jack A. "Characteristics of Unskilled, Skilled, and Highly Skilled Mental Calculators." Unpublished doctoral dissertation, University of British Columbia, 1984.

Josephina, Sister. "Mental Arithmetic in Today's Classroom." *Arithmetic Teacher* 7 (April 1960): 199–200, 207.

Levin, James A. "Estimation Techniques for Arithmetic: Everyday Math and Mathematics Instruction." *Educational Studies in Mathematics* 12 (November 1981): 421–34.

Maier, Eugene. "Folk Math." *Instructor,* February 1977, pp. 84–89, 92.

National Advisory Committee on Mathematical Education (NACOME), Conference Board of the Mathematical Sciences. *Overview and Analysis of School Mathematics, Grades K–12.* Reston, Va.: National Council of Teachers of Mathematics, 1975.

National Assessment of Educational Progress. *The Third National Mathematics Assessment: Results, Trends, and Issues.* Denver: NAEP, 1983.

National Council of Teachers of Mathematics. *An Agenda for Action: Recommendations for School Mathematics of the 1980s.* Reston, Va.: The Council, 1980.

———. *Priorities in School Mathematics: Executive Summary of the PRISM Project.* Reston, Va.: The Council, 1981.

Reys, Robert E. "Mental Computation and Estimation: Past, Present, and Future." *Elementary School Journal* 84 (May 1984): 547–57.

Reys, Robert E., James F. Rybolt, Barbara J. Bestgen, and J. Wendell Wyatt. "Processes Used by Good Computational Estimators." *Journal for Research in Mathematics Education* 13 (May 1982): 183–201.

Robitaille, David F. *Summary Report: British Columbia Mathematics Assessment.* Victoria, B.C.: Ministry of Education, Learning Assessment Branch, 1981.

Robitaille, David F., and James M. Sherrill. *Instructional Practices: British Columbia Mathematics Assessment.* Victoria, B.C.: Ministry of Education, Learning Assessment Branch, 1977.

Trafton, Paul R. "Estimation and Mental Arithmetic: Important Components of Computation." In *Developing Computational Skills,* 1978 Yearbook of the National Council of Teachers of Mathematics, edited by Marilyn Suydam, pp. 196–213. Reston, Va.: The Council, 1978.

Wandt, Edwin, and Gerald W. Brown. "Non-Occupational Uses of Mathematics." *Arithmetic Teacher* 4 (October 1957): 151–54.

5

Approximation as an Arithmetic Process

Peter Hilton
Jean Pedersen

OUR concern in this article is with approximate arithmetic as a part of applied mathematics. Although numbers are used for both counting and measurement, the arithmetic of measurement, as we shall show, is much subtler than the arithmetic of counting. Of course, approximate arithmetic is also used in the arithmetic of counting to provide rough estimates—very importantly, to provide a check on a calculation to be executed mechanically—but the considerations that feature in this article are not relevant to such crude calculations.

INTRODUCTION

Since our purpose is to discuss the arithmetic of approximate calculation, it would be absurd not to preface that discussion with some remarks relating to the circumstances under which such calculation is used and to the tools such calculation requires. Thus we will divide this article into four parts. This introduction sets the stage for the arithmetic that follows; the next section gives us the language of approximate arithmetic; the arithmetical procedures themselves appear in the third section; and their justification occupies the final section.

If we are told that a certain distance AB is 23.62 meters, this means that an accuracy to within the nearest centimeter is claimed for the measurement of AB. If asked for a quarter of the distance AB, it would be absurd to give the answer 5.905 meters, since we certainly cannot guarantee the figure 5 representing the millimeter reading. Thus an exact arithmetical calculation

---

Some of the ideas expressed in this article were first aired in the authors' publication *Fear No More* (Addison-Wesley, 1983). We are grateful to the publishers for their implicit permission to reventilate them here.

is strictly irrelevant to the problem at hand; what we need is an *approximate* calculation—we would probably be content to say that the quarter-distance is between 5.90 meters and 5.91 meters. We shall give several further examples in which a problem is solved with the help of an approximate calculation; in these examples, as in the example above and in examples to be found in other contributions to this yearbook, an accurate calculation would be irrelevant to the solution of the problem or, at best, only a step toward the solution and perhaps not the most efficient step.

Before taking up these examples, let us bring into the open one of the most popular misconceptions about mathematics. According to this mistaken belief, mathematics, being "certain" and "precise," cannot deal in approximate answers and cannot countenance a number of different and equally valid solutions to a problem. Section 3 will be full of precise approximate answers, illustrating the fact that an approximate calculation can lead to a perfectly definite answer to a question, especially a question involving a decision. Moreover, we need hardly emphasize that when making approximations, we exercise choice, and no one choice is to be picked out as the single, unique right answer. Thus if we want the approximate area of a field that is 12.3 meters by 13.7 meters, then we must admit 170, 168, 169, and 168.5 square meters as valid answers. *Which* answer we choose will depend on the purpose, or context, of the calculation; thus, it is vital to take into account the actual question we want to answer in deciding our strategy with regard to an approximate calculation or estimation.

Typical of questions leading to approximate arithmetic are the following: Can I afford it? Which is the better buy? Do I have enough gas? Which car gets the better mileage? Who has the better average? Is the task possible? Since examples of this type are provided, and discussed, in other articles in this yearbook, we shall be content to illustrate just three of these questions.

1. *Can I afford it?* This type of question occurs frequently (indeed, when the economic condition of the less wealthy declines, with increasing frequency) in real life. I have $10.00 with me when I go to the supermarket and I want to buy a large package of detergent ($6.37), a pound of cod ($1.69), and a dozen eggs ($0.85). Do I have enough money? It is not necessary to do a precise calculation to answer this question. Another problem of this kind would be: I receive a raise of $350 a month. I would like to send my daughter to a private school, but the fees are $3000 a year. Can I afford it? The answer here is yes—assuming I was already coping with my budgetary problems before my raise, that I have not overlooked any hidden expenses involved in my daughter going to school, and that I have allowed for increased taxes due to the raise.

2. *Do I have enough gas?* There was a time when we could travel by car without worrying about getting gas when we needed it. At whatever hour of the day or night, whether on the weekend or on a weekday, there would

always be gas stations open. Today this is no longer the case, so that we have to be far more prudent and careful in estimating whether we can complete our journey given the amount of fuel in the tank (not to speak of that other estimate we also must make—can we afford the trip at all!?). Many factors enter into such an approximate calculation, of which obvious ones are our expected speed, the rate of consumption of gas at that speed, the distance to our destination (or to some reliable source of gas), and the amount of fuel in the tank. There may also be considerations of time as well, since we may need to take into account the times of day at which we may expect to find gas stations open. At certain times of crisis there is the additional hazard created by the possible unavailability of gas. We may need to know how many gallons our tank holds, but this may not be necessary information; for we may have recorded the fuel consumption using the tank capacity of our car, rather than the gallon, as our unit of volume. With all the information available—and the information will consist of estimates—we can decide on a rational strategy: should we fill up now even though the tank is nearly three-quarters full? The point we are making is that the answer may be an absolutely definite yes (or no), even though it is based on an approximate calculation employing estimated data.

3. *Who has the better average?* We judge the merits of athletes (and others) by representing their performance as a certain average—for example, a baseball player's batting average is the ratio of the number of hits to the number of times at bat. Though we maintain that averages are used too much and sometimes in very misleading ways, nevertheless, such averages are part of our normal "outlook on life" and often have to be compared. To say that player A's average is better than player B's is to assert that the fraction (or, better, rational number) corresponding to A's average is bigger than that corresponding to B's. Thus we need to be able to take two fractions and say which is the bigger. This does *not* require that we be able—and willing—to subtract fractions, unless we want to say by how much A's average is superior to B's (and, in fact, the *differences* between their averages is not the best way to answer the "how much" question). We can do an approximate calculation to convert the fractions into approximating decimals as a means of deciding which fraction is bigger; or we can use the rule that the fraction A/B is bigger than the fraction C/D if and only if AD is bigger than BC, and then approximate AD and BC. In any case, the question will practically always be answered without having to do a precise calculation.

THE LANGUAGE OF APPROXIMATION

It is an extraordinary triumph of the human intellect that the same number system may be used for both counting and measurement. Nevertheless, we argue that the *arithmetic* of counting is not at all the same as the *arithmetic* of measurement; and that the failure to distinguish between these two arithmetics (which one might call discrete and continuous arithmetic) is responsible for much of our students' confusion and misunderstanding with respect to decimal arithmetic and the arithmetic of our rational number system. For example, in the discrete arithmetic of the rational numbers, $3 = 3.00$; but as measurements, $3\text{ m} \neq 3.00\text{ m}$, since the first measurement indicates accuracy to the nearest meter, while the second indicates accuracy to the nearest centimeter. In developing the continuous arithmetic of measurement and approximation, we have to bring in new, auxiliary concepts that have no place in the abstract world of the rational number system. Some of these concepts we now propose to make explicit.

It may very well happen that we are given a decimal number to a greater degree of accuracy than we require, and we wish to approximate to that number. There are two basic reasons why this situation arises and it is important to distinguish between them. On the one hand, it may be quite unnecessary to record a number, or the result of a calculation, to the degree of accuracy available. We have given examples of this situation in the previous section. Again, if a magazine costs \$2.25 a week and there is an 11 percent sales tax and we want to know if the annual cost exceeds \$120, it is unnecessary to calculate that 11 percent of \$2.25 is exactly 0.2475 dollars—we would "round off," probably to 25 cents, obtaining the figure of \$2.50 for the weekly outlay, and deduce that our \$120 is not quite enough.

On the other hand, there are situations in which the accuracy we have achieved in a calculation is really quite serious. This happens frequently today when we use hand calculators so extensively. A hand calculator gives us answers to at least seven places of decimals, and our measurements will usually not warrant such accuracy. Let us give some examples now that are independent of the hand calculator. If we want to mark out a square plot of area roughly 2 km^2, then the length of a side can be calculated as 1.414 km. But it is absurd to claim accuracy to the nearest meter when we're dealing with a length like 1.4 km, so we would be—and should be—perfectly content to record the length as 1.4 km. Or if we want to know the area of a rectangular field that is 1.26 km by 0.81 km, it would be misleading to announce that the area is 1.0206 km^2, since we know the lengths of the sides only to the nearest dekameter (= 1/100 km). A sensible report of the area would be 1.02 km^2.

Thus there are situations in which we round off in view of the accuracy we *need* and situations in which we round off in view of the accuracy we can genuinely *get*—and, of course, situations in which both kinds of consideration apply. The technique of *rounding off* is a familiar one, but we emphasize that the technique, easy in itself, is based on the nontrivial, nonalgorithmic art of deciding the appropriate level of accuracy.

We now discuss *measures* of level of accuracy, since this appears to be a matter in need of some clarification. The level of accuracy desired may be expressed as a unit measure: "to the nearest cent," "to the nearest millimeter," "to the nearest gram," and so forth. However, if our problem has been translated into arithmetical terms, then the level of accuracy will itself be translated into purely numerical terms. Thus if the unit cost of some item is $12.44 (to the nearest cent, or as a precise amount) and we record the amount in *dollars,* we will obtain the number 12.44 *to the nearest hundredth,* or *to two places of decimals.* Again, if the weight of some object is approximately 3.280 kg (using a very sensitive weighing device!) then we may record the weight in kilograms as 3.280 to the nearest thousandth or to three places of decimals. Of course we might choose to report the weight in *grams* instead of kilograms, in which case the number recorded would be 3280 to the nearest whole number. Sometimes it is more natural to describe the accuracy by the number of significant figures rather than the number of places of decimals; this allows for changes of basic unit, as in the example above. However, we must then be careful to specify, when giving a whole number ending with zeros, such as 24 600, whether the zeros are significant (i.e., reliable) figures. So 24 600 could contain three, four, or five significant figures.

Thus, in a rounding-off problem, the data consist of a set of numbers together with a *level of accuracy* stated in the form "to the nearest hundred," "to the nearest whole number," "to the nearest thousandth," and so on. The technique of rounding off is well known—we find the (whole number) multiple of the stated level of accuracy closest to the number given, and this we do by cutting off digits to the right. There are certain special points to be noted, however.

1. If the piece cut off starts with any of the digits 5, 6, 7, 8, or 9, we must *increase* the approximation by 1 in the last (right-hand) surviving place (thus 23.57 to the nearest whole number is 24, to the nearest tenth is 23.6). There is a small exception to this rule: if what we cut off is *exactly* 5, we may increase or not, according to our choice. Some textbooks give eccentric and unsound variants on this eminently sensible way of dealing with this exception!
2. If we require accuracy to the nearest ten, hundred, thousand, and so on, then zeros will inevitably appear in our round-off approximation.
3. If we round off 17.01, say, to the nearest tenth, we must announce the result as 17.0, *not* 17, for if we suppress the zero after the decimal point, then our answer would be interpreted as approximating to the nearest whole number.
4. If the number N is to be rounded off, it cannot be rounded off to an accuracy greater than that to which the number N is itself given. Round off 26.3 to the nearest tenth—answer, 26.3. Round off 26.3 to the nearest hundredth—answer, IMPOSSIBLE!

So far we have discussed rounding off to the *nearest* tenth, hundredth, ten, and so forth. However, there are situations when we need to round *up* or round *down.* For example, if a supermarket is selling cans of sardines at $1.61 for two cans, then you may be sure that you would have to pay 81¢, not 80¢, for one can. Thus in working out unit prices where the price is quoted in bulk, we always round *up.* You have probably noticed that in the marketplace this rounding up happens even when the number determined by the appropriate arithmetic is really closer to the lower number. Thus if apples are priced at 4 for 61¢, then even though 61 divided by 4 is 15.25 (certainly closer to 15 than 16), you would be charged 16¢ for one apple.

If, on the other hand, we are told to buy 6 items of a certain kind and not to spend more than $10, then, to determine the maximum unit price we can afford, we carry out the required division, obtaining an answer (if we use a hand calculator) of 1.6666667, and we *must* round down to $1.66. The context of an approximation will determine whether we will need to round up, down, or off. If we do need to round *down,* it should be clear that we simply carry out the process outlined above for rounding off, except that our special point 1 above should be ignored. If we need to round *up,* we modify the process by always increasing the approximation by 1 whenever there are nonzero digits in the piece cut off. Thus, rounding down 27.681 to the nearest tenth, we get 27.6; rounding up 161.24 to the nearest whole number, we get 162. Remember, then, that we speak of three different kinds of rounding process: rounding *off,* rounding *up,* and rounding *down.*

We shall be discussing in the next section the special arithmetic that applies to computations that involve various combinations of rounded numbers. Here, however, we shall take up a very important question: is there a way of presenting approximate numbers that facilitates their understanding and their arithmetic? Scientists and engineers have indeed arrived at an agreed opinion on this—the result is what we call *scientific notation.* If, for example, we wish to present the approximation 32.58 in scientific notation, we rewrite the number with the decimal point following the first digit on the left—this is, we write 3.258; and then we "adjust" by multiplying by a suitable power of 10, in this case, 10^1. Thus, in scientific notation,

$$32.58 = 3.258 \times 10^1 \ (= 3.258 \times 10).$$

Similarly,

$$3987 = 3.987 \times 10^3,$$

$$46\,910\,000 = 4.691 \times 10^7.$$

(The student should notice that the power of 10 turns out to be just the number of places we have moved the decimal point to the left.)

We now note that to use scientific notation for numbers less than 1, it is necessary to have available negative exponents. Thus the development of the appropriate arithmetic will have as prerequisite a familiarity with the full arithmetic of the integers, negative as well as positive, even if we are dealing only with the approximate arithmetic of *positive* numbers. We should add that the use of negative exponents provides a vivid, and entirely valid, justification of the vexatious rule

$$(-a)(-b) = ab.$$

Having introduced scientific notation, we must meet the question, what are its advantages? We claim that an expression like 4.691×10^7 is easier to grasp and manipulate than 46 910 000. Having the exponent 7 presented as part of the notation enables us to see immediately the *order of magnitude* of the number. We have met people in this country who could not remember whether the defense budget was \$126 million or \$126 billion. (The encounter was in 1979. The defense budget is now more than twice the figure of \$126 000 000 000.) Thus they remembered the relatively insignificant figure 126 but failed to remember the more significant order of magnitude. We contend that had they been familiar with scientific notation, it is very unlikely that they would have confused 1.26×10^8 with 1.26×10^{11}. Moreover, scientific notation makes it considerably easier to make comparisons between numbers. Thus if the defense budget is regarded as sacred—indeed, it is so sacred it is constantly increasing—it is easy to see why we have been so unsuccessful in our attempts to reduce government spending. Whenever we make inroads in, say, our social services, we are dealing with expenditures measured in millions (10^6), while the defense budget is measured in billions ($10^9 = 10^3 \times 10^6 = 1000 \times 10^6$).

Thus in scientific notation we exhibit the *order of magnitude* (10^{11} in our defense budget example) and the *significant figures* (1.26 in that same example). Multiplication and division of approximate numbers are also facilitated by this notation, since we multiply or divide powers of 10 by using the law of exponents, and we may easily estimate the result of multiplying or dividing numbers between 1 and 9, thus providing a check on a computation on our hand calculator. We shall give examples of these concepts in the next section.

We are certainly *not* suggesting that every time you make an approximation or an estimate you must record it in scientific notation. Scientific notation should be used for quantities you may need to remember, especially if they are very large or very small, or for quantities on the basis of which you wish to make further estimates, using the appropriate arithmetic. If it is just a matter of some transitory interest, some approximate calculation that, once done, does not need to be recorded in one's memory or used for some further arithmetical purpose, then by all means one should feel free to keep it in the form in which it naturally arose. We should always use the notation best adapted to our purpose.

THE ARITHMETIC OF APPROXIMATION

The arithmetic of measurements and approximations is, of course, based on the familiar arithmetic, with respect to concepts, operations, and algorithms, but it is different because its purpose is different. We have not been impressed by the integrity of some of the presentations of this subject that we have seen, where what is dignified by the name "approximate arithmetic" is merely the approach of the lazy person to traditional arithmetic. That is, perfectly definite decimal numbers are given and the student is invited to perform the usual arithmetical operations on these numbers but to present the answer to some specified but arbitrary accuracy. In the real-world uses of the arithmetic of approximation, the numerical data are usually uncertain, so that the outcome of the arithmetic is also uncertain and a key question to answer is, how uncertain? This is a delicate issue (far more delicate for multiplication and division than for addition and subtraction). We begin with an illustrative example involving subtraction.

Suppose the odometer on our bicycle, calibrated in tenths of a kilometer, reads 263.6 when we leave the house, and 287.2 when we arrive home again. How far have we cycled? The obvious answer is 287.2 − 263.6, or 23.6 km, but can we be sure of the accuracy of the last digit, 6? Let us study the situation a little more closely. The reading 263.6 actually means that the bicycle had been ridden between 263.6 and 263.7 kilometers when we started out; and the reading 287.2 means that the bicycle had been ridden between 287.2 and 287.3 kilometers when we returned. Thus the shortest distance we might have traveled is 287.2 − 263.7 = 23.5 km, and the longest distance we might have traveled is 287.3 − 263.6 = 23.7 km. We cannot be sure we have ridden 23.6 km (to the nearest tenth of a kilometer); we might have ridden anything from 23.5 to 23.7 kilometers. It is easy to see (we shall present a generalized argument later) that 23.6 remains the *most likely* estimate (to one place of decimals) of the distance we have traveled, being much more likely than 23.5 or 23.7, which are equally probable. How then should we answer the question, "How far did you cycle today?" The best answer would be, "23.6 km, but it could be 23.5 or 23.7, to the nearest tenth of a kilometer." This kind of situation, and the calculations involved, are typical of a subtraction problem based on measurements that have been rounded down (as with the odometer reading) or rounded up, whereas we wish to present the answer rounded off. Let us now describe an addition problem in which the addends are rounded up and it is both natural and correct in the context to give the answer rounded up also.

Suppose there is a regulation in a school district that on any outing there must be an adult to accompany every ten children. Thus, for example, if 83 children go on an outing, they must be accompanied by 9 adults (rounding 83/10, or 8.3, up to 9). Now let us assume that if the children from school A go on an outing, 12 adults are required; and if the children from school B go on the outing, 14 adults are required. How many adults would be required if, instead, schools A and B combined for the outing? To obtain the answer, we reason as follows. We know that there are between 111 and 120 children from school A and between 131 and 140 children from school B going on the outing. If the schools combined, between 242 and 260 children would go on the outing. Thus we may need 25 or 26 adults. Which number is more likely? It turns out (see the next section) that the probability that we would need 25 adults is 9/20 (= .45) and the probability that we need 26 adults is 11/20 (= .55). These probabilities of 9/20 and 11/20 are independent of the numbers 12 and 14 in the data of our problem—but this may not be clear immediately to the student. Our conclusion is that, with this regulation, we would expect to save an adult's time (and the school district's money) nine times out of twenty by combining two schools when going on outings.

What, then, are the factors involved in deciding how to do an addition or subtraction problem involving approximations? First, we must realize that we have different rules for addition problems and subtraction problems. Second, we must take account of which way our data are rounded and which way our answer should be rounded. Having recognized these aspects, we begin to realize that the arithmetic of approximations is rather subtle, requiring us to proceed carefully and systematically.

We shall give the rules covering various common situations after discussing a third example. Suppose we are rounding off to the nearest hundredth and have to add the approximations 17.24 and 8.37 (this problem might well arise in adding lengths measured in centimeters and recorded in meters). Then the obvious answer of 25.61 is the most likely, but the actual answer might be 25.60 or 25.62. For example, the measurements might more accurately have been recorded as 17.236 and 8.366, when the sum would plainly be 25.60 to the nearest hundredth; or they may have been recorded as 17.244 and 8.373, when the sum would be 25.62, to the nearest hundredth.

In giving the rules that follow, we shall indicate the general principle by means of a typical special case; moreover, we shall always round to *two* decimal places, trusting to the reader to make the necessary and obvious change in the rules if we are rounding to a different degree of accuracy. If the solution to a problem could be 25.51, 25.52, or 25.53, we shall represent this as 25.52 ± 0.01. It should then be understood that 25.52 is the most likely answer.

Of course, we are not presenting this material as it should be presented to students; we are merely summarizing the key results. But we believe that our own pedagogical principles are clear from our examples and from the style of this article. Naturally, the student should try to understand how the results we are about to present are obtained.

First, then, we shall deal with *addition*. There are five important cases, and they may be classified as follows:

Rules for Adding Approximations

Case	Data rounded	Solution rounded	Sample problem	Solution
1	Up	Off	12.81 + 3.42	16.22 ± 0.01
2	Off	Off	12.81 + 3.42	16.23 ± 0.01
3	Down	Off	12.81 + 3.42	16.24 ± 0.01
4	Up	Up	12.81 + 3.42	16.23 or 16.22
5	Down	Down	12.81 + 3.42	16.23 or 16.24

Next we display the appropriate results for *subtraction*. Here there are seven important cases, which we classify as follows:

Rules for Subtracting Approximations

Case	Data rounded	Solution rounded	Sample problem	Solution
1	Up	Off	9.31 − 6.42	2.89 ± 0.01
2	Off	Off	9.31 − 6.42	2.89 ± 0.01
3	Down	Off	9.31 − 6.42	2.89 ± 0.01
4	Up	Up	9.31 − 6.42	2.89 or 2.90
5	Down	Down	9.31 − 6.42	2.89 or 2.88
6	Up-down	Up	9.31 − 6.42	2.89 or 2.88
7	Down-up	Down	9.31 − 6.42	2.89 or 2.90

(In the first five cases, as in addition, we round both numbers in the subtraction problem the same way.)

Let us exemplify the last two cases, which are special to subtraction. Suppose we know the approximate length, a, of our total journey and the approximate distance we have already traveled, b. There are circumstances in which we might seek a *minimum* value for the remaining distance, $a - b$ (we might want to phone friends to tell them not to expect us before a certain time); and there are circumstances in which we might seek a *maximum* value for the remaining distance, $a - b$ (do we have enough gas to complete the journey?). The minimum problem is case (7); we should round a down and b up. The maximum problem is case (6); we should round a up and b down.

You will notice that case 6(7) for subtraction is deducible from case 4(5) for addition. Indeed, more generally, it is still possible to say that, in a sophisticated way, subtraction is inverse to addition.

In the presentation of this material, the following points must be brought out:

1. Handling approximations and estimates is not easy and does *not* follow immediately from the rules of elementary arithmetic.
2. There are many different situations to be dealt with, and the particular situation depends not on our choice but on the nature of the data and the nature of the problem.
3. The most obvious answer is reasonable but uncertain, and in any case, in the arithmetic of approximations, *the answer is less certain and less accurate than the data.*

At this point we recall an important remark from the introduction. For certain approximate calculations, and most especially when doing a rough calculation merely to check an accurate machine calculation, it will not matter how you rounded the data or how you should round the answer. You will be content with a "rough estimate," and you will not need to know just how accurate your estimate and your answer are. *Nothing we have said in this section applies to such a situation.* For example, you may well find it convenient, in a rough addition of this kind, to round one addend up and the other addend down. Here we are really concerned with solving real-world problems, in which the nature of the problem, and of the measuring instruments used, naturally determines how the data are rounded and how the solution should be rounded.

We turn next to *multiplication.* To develop a precise set of rules such as we have given for addition and subtraction would involve us in excessive complication that would not be justified by its practical advantages. Moreover it is to be expected that most multiplications will be done on a hand calculator, so that it will be largely a matter of having a reasonable sense of the degree of accuracy and reliability of our answers. Thus we shall concentrate on the technique of approximate multiplication and the reliability of the answer.

The precise product of 21.2 and 4.7 is 99.64. However, if we are told that a plot of land is 21.2 m by 4.7 m, it would be absurd to say that its area is 99.64 m^2, since we obviously can put no faith in the digit 4 in the second place of decimals. But, in fact, the answer is less reliable than even that remark suggests. For if the plot had measured 21.15 m by 4.65 m, its area would be about 98.3 m^2 (since the precise calculation yields 98.3475); and if it had measured 21.25 m by 4.75 m, its area would be about 100.9 m^2 (the precise calculation giving 100.9375). Thus the figure we have obtained (rounding off to the nearest tenth) of 99.6 m^2 is the value of the midpoint between the two extreme possibilities (the average). Moreover, the spread of possible values around the midpoint 99.6 m^2 is ± 1.3 m^2, and

$$1.3 = .05 \times (21.2 + 4.7),$$

where .05 is the possible error in our measurements of length (since we are measuring to the nearest tenth) and $21.2 + 4.7$ is the sum of the lengths. (See fig. 5.1.)

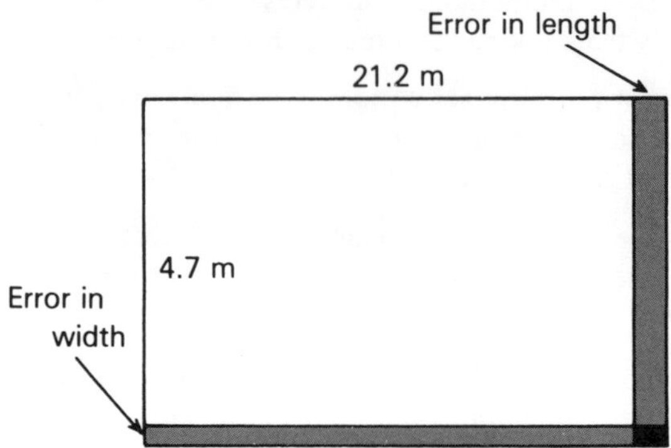

Fig. 5.1. Visualizing the error in an area calculation

This exemplifies a very good approximate rule, whose validity we shall establish in the final section. Thus, if we wish to multiply two numbers, each rounded off to the same given degree of accuracy, we multiply on a hand calculator and round off the answer to the same degree of accuracy. This is then our *best estimate,* but it is subject to an error of

$$\pm \text{(maximum error in our measurements)} \times \text{(sum of the numbers)}.$$

Let us give another example of this rule. If the density of a substance is 9.11 (mass per unit volume) and if the volume is 12.35 (measured in some appropriate unit, say, cm^3), then its mass (measured in appropriate units) is approximately 9.11×12.35. From the hand calculator we obtain the estimate 112.51. The possible error, calculated from the rule above, is

$$\pm (.005) \times (9.11 + 12.35) = \pm (.005) \times (21.46) = \pm 0.11.$$

Frequently estimates to be multiplied will be presented to us in scientific notation. For example, we may be required to find the product

$$(3.1 \times 10^5) \times (4.6 \times 10^2).$$

We proceed as above to find the product 3.1×4.6, which is 14.3 ± 0.4. Then we use the law of exponents to complete the multiplication. Thus we obtain the central estimate

$$(3.1 \times 10^5) \times (4.6 \times 10^2) = 14.3 \times 10^7.$$

There are two important things to notice about this estimate. The first is that the possible error of ± 0.4 in the calculation that gave us 14.3 should also be multiplied by 10^7 to give the possible error in our estimate. Thus our answer may be larger or smaller than 14.3×10^7 by an amount of 0.4×10^7. The second thing to notice is that the estimate given is not, in fact, in scientific notation, since 14.3 does not lie between 1 and 10. Thus to put the estimate in scientific notation we should rewrite the answer as

$$1.43 \times 10^8, \text{ with a possible error of } 4 \times 10^6.$$

As a further example, involving negative exponents, we have

$$(1.25 \times 10^2) \times (7.20 \times 10^{-2}) = 9.00 \times 10^0 = 9.00,$$

with a possible error of 4×10^{-2}.

We now make two further remarks, relevant to real-world arithmetic.

Remarks

1. In addition and subtraction situations, the two quantities being added or subtracted are always of the *same kind.* We add (or subtract) lengths, we add (or subtract) sums of money; we do not add a length to a weight, nor do we subtract an area from a volume. Thus it is natural that, in such situations, the two numbers involved be given to the same level of accuracy. However, in multiplication this is often not so. We may indeed multiply quantities of the same kind—as, for example, when we multiply lengths to obtain areas. But very often we have to multiply numbers representing very different quantities. Thus we may multiply speed by time to obtain distance traveled; we may multiply unit cost by quantity purchased to obtain total cost; we may multiply mass by acceleration to calculate force. In these cases there is, of course, no reason why the measurements made should be expressed by decimal numbers recorded to the same degree of accuracy—it is, indeed, quite meaningless to ask for the *same* degree of accuracy when *different* measures are involved. There are rules governing the uncertainty involved when this situation arises, but they are more complicated than the rule given above and we are not going to insist on them now. (They are appropriate to a calculus course, but they will be given in the next section!) As a practical rule we recommend that if a real-life problem gives rise to a multiplication of two approximate measures, you should round off your answer (obtained on a hand calculator, probably) to the lesser accuracy of the two numbers involved and simply announce that as your best estimate. Thus, purchasing 1260 articles (to the nearest 10) at \$0.314 each (the unit price being given to the nearest thousandth of a dollar, or *mil*), we find $1260 \times 0.314 = 395.640$, so the total cost should be announced as \$400, as closely as one can say.

2. As a special case of the situation discussed above there are many situations in which we have to multiply an estimate or an approximate amount by an *exact* whole number. We may, in other words, be using multiplication by a whole number as repeated addition, and we will often then know precisely how many times to take our estimated amount. If the time of sunset advances approximately 613 seconds each day in May, it will advance approximately $7 \times 613 = 4291$ seconds in a week and we are not

going to approximate to 7 to do this calculation! How then should we handle this type of question? Here the exact calculation gives the best estimate—that is clear; but how far out might we be? The answer is very simple. If we are multiplying X by N, where X is a measure to the nearest hundredth, say, and N is an exact whole number, then our best estimate is NX and our possible error is $\pm N/200$, since the possible error in X is $\pm 1/200$.

We do not wish to say too much about *division* here, since division is a very delicate subject, and we are constrained by the maximum length permitted to this article (emphatically, in approximate arithmetic, division is not the inverse of multiplication). However, we can give a reasonable rule for dividing one approximation by another. If both numbers are given to the same degree of accuracy—this presupposes that both numbers represent measurements of the same kind, so that their ratio is a *number* (not a length, a weight, a time interval, and so on)—then we may divide one by the other on our hand calculator and round off our answer either (a) to the nearest whole number if the nature of the problem forces the answer to be a whole number or (b) to the same degree of accuracy as our original measurements if a decimal answer is sensible. (We are never justified in giving an answer to such a division problem to more significant figures than are given in the divisor, contrary to traditional textbook practice.) Of course, it is understood in alternative (a) above that our original measurements were themselves at least as accurate as to the nearest whole number. Let us now illustrate what we have said so far with two examples before proceeding to discuss the question of possible error.

If we know that the weight of an individual tack is 23 grams and we have a quantity of tacks weighing 20 120 grams, then our best estimate for the number of tacks is 875; it would be absurd to announce an answer of 874.8, since we must have a whole number of tacks! If, on the other hand, we are told that the atomic weights in appropriate units (the weight of a hydrogen atom, say) of two chemical elements are 94 and 91, then the ratio of their weights may be given as 1.03.

Now for the possible error. If we have to divide X by Y and each of these quantities is given, say, to three places of decimals, so that the possible errors in X and Y are $\pm h$, where $h = 0.0005$, then, as we show in the final section, the possible error in X/Y is

$$\pm \frac{h(X + Y)}{Y^2}.$$

Note that this error may be dangerously large if Y is very small compared with X—and this is a rather common situation. For example, if our measurements of weight above were accurate to the nearest gram, then $X = 20\,120$, $Y = 23$, and $h = 0.5$, so that, using the formula above, the possible error is

$$\frac{\pm 0.5 \times (20\,120 + 23)}{23^2} = \pm\frac{0.5 \times 20\,143}{529}$$
$$= \pm 19.04, \text{ approximately.}$$

If we are dealing with a division situation in which X and Y represent quantities of a quite different kind (for example, distance and time)—so that their ratio is not a "pure number," as in the case of the weights above—then the best practical rule we can give is to accept no answer containing more significant figures than there are in the divisor. (Of course, we may be required in the statement of the problem only to make a crude estimate of the quotient, but here we are assuming that it is up to us to choose the appropriate level of accuracy.) A warning is necessary at this point, just as in our second remark relating to multiplication: namely, the divisor may well be an exact and precise whole number. In such a case, of course, the practical rule relating to significant figures in the divisor is no help to us, and we must be guided by the actual problem itself, the purpose of the calculation, and, in the final resort, by the degree of accuracy of the dividend.

All that we have been doing in this section is to suggest some helpful rules. We have very deliberately been quite undogmatic. The strong impression we hope to have left with the reader is that when an approximate calculation is done, the "solution" is itself to be regarded as an estimate—and, generally speaking, not as reliable an estimate as those that formed the data for the calculation.

MATHEMATICAL DERIVATIONS

We now take up four mathematical questions referred to in the previous section. First we consider a real-variable problem.

THEOREM 1. *Let x, y be random variables distributed with equal likelihood in the range $0 \le x,y < 1$. Let $N = N(x,y)$ be the integer nearest to $x + y$. Then*

$$\text{prob}\,\{N = 1\} = \frac{3}{4},$$

$$\text{prob}\,\{N = 0\} = \text{prob}\,\{N = 2\} = \frac{1}{8}.$$

Proof. Study the diagram in figure 5.2.

This theorem, reinforced by obvious scaling devices, establishes the probabilities of the three possible solutions given in our rules for adding and subtracting approximations when the solution is rounded off.

We next consider a discrete problem.

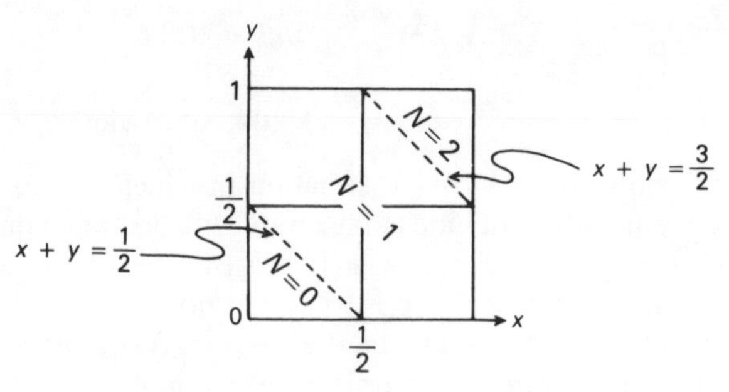

Fig. 5.2

THEOREM 2. *Let X, Y be random variables taking the values* 1, 2, . . . , N *with equal likelihood. Then*

$$\text{prob}\,\{X + Y \leq N\} = \frac{N - 1}{2N}.$$

Proof. Consider the possibilities $X + Y < N + 1$, $X + Y = N + 1$, $X + Y > N + 1$. Let their probabilities be p, q, r. By symmetry $p = r$ and, of course, $p + q + r = 1$. It is easy to see that

$$q = \frac{N}{N^2} = \frac{1}{N}, \text{ so that } p = \frac{1}{2}(1 - q) = \frac{N - 1}{2N}.$$

This theorem, together with the evident companion theorem relating to $X - Y$, establishes the probabilities of the solutions in our rules for adding and subtracting approximations when the solution is rounded up or down. (Look at our problem of the two schools merging for the school outing, where $N = 10$.)

We now turn to multiplication and division. We were concerned in the previous section with the question of the reliability of the estimate AB of the product of two numbers whose estimated values are A and B. We also raised a similar question related to division.

Let us suppose then that A is subject to an error of $\pm h$ and that B is subject to an error of $\pm k$. We assume that h and k are positive and small compared with A and B. Then the biggest value the product can have is $(A + h)(B + k)$, and the smallest value is $(A - h)(B - k)$. Now

$$(A + h)(B + k) = AB + kA + hB + hk.$$

However, hk can be neglected as being very small compared with kA and hB, so that the greatest possible value is, effectively,

$$AB + kA + hB.$$

Similarly the smallest possible value is, effectively,

$$AB - kA - hB.$$

We conclude that the estimate AB is subject to the error

$$\pm(kA + hB).$$

We drew particular attention in the previous section to the case $k = h$. Then the error term may be seen to be $\pm h(A + B)$. This is precisely the rule we gave.

Finally, we pass to the division problem. We suppose that X is subject to an error of $\pm h$ and that Y is subject to an error of $\pm k$, and we again assume that h and k are positive and small compared with X and Y. We seek to measure the possible error in the quotient when X is divided by Y. Now the greatest value the quotient could have is

$$\frac{X + h}{Y - k}$$

and

$$\frac{X + h}{Y - k} - \frac{X}{Y} = \frac{kX + hY}{Y(Y - k)};$$

moreover,

$$\frac{1}{Y(Y - k)} - \frac{1}{Y^2} = \frac{k}{Y^2(Y - k)}.$$

Thus this difference is small if k is small, so we may say that

$$\frac{X + h}{Y - k} - \frac{X}{Y}$$

is, effectively,

$$\frac{kX + hy}{Y^2} \text{ (replacing } \frac{1}{Y(Y - k)} \text{ by } \frac{1}{Y^2}\text{)}.$$

Similarly, the smallest value the quotient could have is

$$\frac{X - h}{Y + k}$$

and

$$\frac{X}{Y} - \frac{X - h}{Y + k} \text{ is, effectively, } \frac{kX + hY}{Y^2}.$$

We conclude that the estimate X/Y is subject to the error

$$\pm \frac{kX + hY}{Y^2},$$

just as we described in the previous section in the special case $h = k$. Notice that there is another important special case, namely, when there is no possible error in Y (typically, Y is a whole number). Then $k = 0$, so that the estimate X/Y is subject to the error $\pm h/Y$. Of course, this case is particularly easy, and there was no need for the elaborate mathematical reasoning above to establish it.

Benchmark: Number Patterns and the Development of Expert Mental Calculation

John A. Hope

MOST great mental calculators report that their methods of calculation were largely self-taught and discovered through playful exploration of some interesting number pattern. For example, Arthur Benjamin, a modern-day calculating prodigy, discovered the method used by most experts to square numbers when he was still a youngster in elementary school. To while away the time spent on a bus, he thought about numbers and noticed the following curious pattern about the products of two numbers whose sum is 20:

$$
\begin{aligned}
9 \times 11 &= 100 - 1 = 10^2 - 1^2 \\
8 \times 12 &= 100 - 4 = 10^2 - 2^2 \\
7 \times 13 &= 100 - 9 = 10^2 - 3^2 \\
6 \times 14 &= 100 - 16 = 10^2 - 4^2 \\
5 \times 15 &= 100 - 25 = 10^2 - 5^2 \\
4 \times 16 &= 100 - 36 = 10^2 - 6^2 \\
3 \times 17 &= 100 - 49 = 10^2 - 7^2 \\
2 \times 18 &= 100 - 64 = 10^2 - 8^2 \\
1 \times 19 &= 100 - 81 = 10^2 - 9^2
\end{aligned}
$$

He soon realized through further trial-and-error experimentation with other numbers that this pattern could be used to square large numbers. For example, to square 37, he used 40 (an easy number to multiply), which is $37 + 3$. Then, since $34 = 37 - 3$, he noted that $37^2 = 40 \times 34 + 9 = 1360 + 9 = 1369$. Through continual practice, he soon was able to recall the squares for virtually all two-digit numbers. Now he can square even large numbers, such as 4273, by successive applications of this method. Not until much later did he understand that this method is simply a recasting of the algebraic identity for the difference of squares, namely, $a^2 = (a + b)(a - b) + b^2$.

In a recent study of the reasoning used by fourth-grade children to calculate the basic facts of multiplication, a young girl explained that she calculated 7×9 by thinking "$64 - 1 = 63$." Further questioning revealed that she squared the middle number and subtracted 1 to calculate products whose factors differed by 2: for example, 7×9 would be calculated as

$8 \times 8 - 1$. This multiplication shortcut was discovered by comparing squares with the products of numbers adjacent to the squares listed in a multiplication table:

×	0	1	2	3	4	5	6	7	8	9
0	0									
1		1								
2	0		4							
3		3		9						
4			8		16					
5				15		25				
6					24		36			
7						35		49		
8							48		64	
9								63		81

Another girl discovered this same pattern in grade 4 and now at the age of thirteen calculates the products of large numbers such as 49×51 by thinking $50^2 - 1$. Neither child knew the algebraic basis of this pattern, namely, $(a + 1)(a - 1) = a^2 - 1$.

George Parker Bidder, a famous nineteenth-century mental calculator, believed that his calculating powers had very humble origins. As a young child, he arranged concrete materials such as peas and marbles into rectangular arrays and began to recall many basic facts of multiplication well before he knew the word *multiply*. He noticed that larger products could be calculated by a clever rearrangement of a larger array into smaller arrays. For example, a 13×17 array would be rearranged into a 10×10, 10×7, 10×3, and 3×7 array. The patterns suggested by the physical rearrangement of concrete materials developed eventually into a mental process that Bidder used to calculate the products of six-digit numbers with ease.

Number patterns have fascinated people for centuries. For many expert mental calculators, they not only are fascinating but are, in fact, the likely key to their incredible calculating skill.

6

Linking Estimation to Psychological Variables in the Early Years

Heather L. Carter

About how long will it take to make your presentation to the school board? About how much paint do you need for the fifth-grade project? About how many tapes do you need for recording the interviews? About how many people do you anticipate will attend the reception for the retiring staff member? These are typical of the questions we answer with estimates, but how do we learn to estimate?

Reys (1984) suggests that adults who have acquired the ability to make reasonable estimates have generally taught themselves the skill. Little systematic attention has been directed toward either teaching estimation in schools (Buchanan 1978) or conducting research on the topic. Reys suggests that a more organized approach is necessary. Some materials for teaching estimation have been prepared (Reys, Trafton, Reys, and Zawojewski 1984; Schoen, Friesen, Jarrett, and Urbatsch 1981). These instructional materials, however, are limited to computational estimation and are directed to pupils above the second-grade level. That target seems reasonable in that computational estimation is hardly relevant until students can deal with problems involving two-digit numbers or three addends. Estimation, however, is not limited to situations involving computation. The skill to make estimates of measure and numerosity can and should be developed in the primary grades. It is to these educational levels that we will turn our attention.

When developing instructional materials for estimation, it is important for teachers to consider psychological variables that might be related to the ability to estimate. These relationships have received scant consideration. Two notable exceptions are the works of Reys, Rybolt, Bestgen, and Wyatt (1982) and Siegel, Goldsmith, and Madson (1982). However, both of these works deal with postprimary pupils.

Three major psychological variables that are logically related to the ability to estimate need to be considered in the early school years. These are *error tolerance, cognitive structuring,* and *decomposition/recomposition.* The relationship between each of these variables and estimation will be posited, and implications for teaching and research will be discussed.

ERROR TOLERANCE

A tolerance for error is important if children are to learn that an approximate answer may not only be acceptable but is, in fact, more appropriate in certain situations than an exact answer. The importance of error tolerance in estimation is supported by Reys and his colleagues (1982, 197). After completing a computational estimation test, superior estimators from grade 7 to adult were identified and interviewed to determine any common solution strategies or common characteristics. All the good estimators were found to have a tolerance for error.

Although very young children have a high tolerance for gross estimations, they quickly become convinced that an exact answer is the only one appropriate to any mathematical question or problem. There does not appear to be an intermediate phase in which children retain their tolerance for error while developing the ability to hone their estimates. Yet this is precisely what is needed if children are to become good estimators.

Little attention is given in the primary grades to problems that might enhance a tolerance for error, such as problems for which exact answers are virtually impossible to determine or for which the only reasonable answers are estimates. Two procedures described by Siegel and colleagues (1982) seem relevant for teaching children how to solve such problems—the act of guessing and the use of benchmarks, or cues. Both involve tolerance for error.

Guessing

"Don't just guess!" is a statement that teachers make repeatedly. Although it is appropriate in some situations, it certainly is not in others.

At times people are forced to guess because they have no other means for arriving at an answer. A person not accustomed to seeing large flocks of geese might say, "There must be two or three hundred geese out there by the lake." It is likely that the number would not be very accurate—it merely indicates a large flock. Quite young children also experience many instances when guessing is the only strategy they have available to them:

- How many bees are at work in the glass-backed hive?
- How far is it to your grandmother's house?
- How long will it be before your baby brother learns to walk?

- How high is the air balloon?
- How tall is the building?
- How many people are in the football stadium?

Although adults might have the background to answer some of these questions, or at least to give reasonable estimates, most primary school children do not. They need to be able to say, "I just guessed," and to receive the answer, "That's O.K."

Informal situations in the classroom can be used to encourage guessing and at the same time lay the groundwork for more sophisticated estimation procedures. For instance, when trees are being studied, a discussion might proceed as follows: "I wonder how many leaves there are on the tree. How many do you think? I will record your answers." "Lots. . . . A million. . . . Two million. . . . A hundred." "Why is it difficult to get an accurate answer to the question?" "Because there are so many leaves. . . . They can't be counted. . . . We have to guess." "Well, let's try a small twig." "Twenty. . . . Fourteen. . . . Eight. . . . Ten." "Is that easier? Why?" "Because it's smaller. . . . Because we can almost count them." Situations such as these help to develop two abilities in children: (1) the ability to recognize situations in which guessing is the only reasonable strategy even though it provides no more than a gross estimate; (2) the ability to recognize that different levels of accuracy are possible and acceptable in different situations. Both abilities are prerequisite to becoming a good estimator, and both require a tolerance for error.

Using Benchmarks

A slightly more sophisticated step in the development of estimation skills is the ability to use benchmarks, or cues, as described by Siegel and colleagues (1982). Benchmarks are nonstandard units such as familiar objects or events that the child can use as referents in order to estimate. The use of benchmarks may lead to more accurate estimates than simple guessing, but the child still needs a tolerance for error in order to accept the estimate as a reasonable solution.

Frequently, situations arise in which children ask such questions as "How long will it take to get to the art gallery?" or "How far is it to the public library?" To answer the first question with "45 minutes" or the second question with "three-quarters of a kilometer" probably makes little sense to young children. To say, "About as long as it takes to get to the zoo" or "About as far as the park" is more meaningful. These are relevant benchmarks for young children.

Children's facility with these benchmarks will be enhanced if informal discussions follow, such as "What else takes about the same time?" and "What would take a little more time or a little less time?" Later the children will reach the stage where they are able to say, "About twice as far as" or "About half the time." They will thus begin to integrate mathematical operations and estimation.

To become good estimators, children need to recognize situations where benchmarks are available and then use them to generate reasonable estimates. Developing the ability to refine their estimates can then begin, but even at this stage a reasonable tolerance for error is required.

COGNITIVE STRUCTURING

Two closely related cognitive-structuring constructs will be considered: *centering* and *field dependence/independence.* Both constructs relate to the manner in which children view parts of their environment with respect to the whole. Since estimation often involves considering a whole in terms of its parts, these variables appear to be logically related to estimation ability.

Centering

In their early years, young children tend to *center*—that is, focus on only one attribute, or characteristic, of an object or event at any one time (Piaget 1965). To make reasonable estimates, it is frequently necessary to consider multiple attributes simultaneously. For example, an adult, when estimating the number of balls in a jar, will consider several attributes concurrently—the diameter of the jar, the height of the jar, the dimensions of the balls. However, a young child who is centering is able to consider only one attribute at a time—such as the "look" of the balls. This tendency to center will limit the young child's ability to reach a reasonable estimate.

Although it is an unresolved issue whether one can "teach" children not to center, there is nothing to lose—and possibly much to gain—by making an attempt to help them move out of this centering stage and consider multiple relevant attributes. Activities using attribute blocks may well assist this development. The children can initially sort the blocks on the basis of one attribute, later two, and then three. They can play games that require oral descriptions using first one attribute and later multiple ones.

Considering multiple attributes will permit children to take different aspects of a situation into account and thus help them arrive at an estimate. In fact, the development of this ability to consider multiple attributes appears to be essential for efficient estimation strategies.

Field Dependence/Independence

A second cognitive-structuring construct that may be relevant when children estimate is the ability to extract and focus on only the critical elements in a given setting. Witkin, Moore, Goodenough, and Cox (1977) have suggested that this focusing skill depends on the extent to which an individual processes information in a field-independent/dependent manner. Persons are said to be field-independent who focus on critical elements in a situation while discounting, although recognizing, the less salient elements.

The extent of field independence may help explain the variation among adults in their ability to make reasonable estimates in real-life situations. Take, for instance, estimating the real value of a new house. The person who focuses on the basic structure, design, location, and adaptability of the house will probably reach a fairly accurate estimate of the value. The one who takes into account only the "gingerbread items," such as the security system, color of appliances, and built-in outlets for specialized equipment, will probably make a less accurate estimate. The former focused on the critical elements, that is, was field-independent, whereas the latter was influenced by distractors (field-dependent).

Even for primary school children, the skill of focusing on critical factors is important in making estimates. For example, those who take into account the color or weight of the balls in a jar are being distracted by irrelevant attributes, and this is quite likely to interfere with their ability to make a reasonable estimate.

Thus, the ability to estimate appears to be related to field independence as well as to centering. To become good estimators, children need to develop the ability to focus on multiple attributes simultaneously and to select critical elements from a field of less salient ones. By doing both, they increase the likelihood of reaching a reasonable estimate.

DECOMPOSITION/RECOMPOSITION

A major portion of the model of estimation proposed by Siegel and colleagues (1982) is the psychological variable, or process, called *decomposition/recomposition*. Decomposition/recomposition is the process by which the to-be-estimated item (TBE) (p. 213) is first decomposed into samples to which basic estimation skills can be applied and then recomposed to determine the final estimate. For example, an adult uses this process when estimating the number of words on one page of a manuscript. First the TBE (the entire page of words) is decomposed into samples (separate lines). The number of words in a sample (one lines is determined. Recomposition occurs when the sample is used to estimate the TBE (the number of words on the entire page).

This skill of decomposition/recomposition is somewhat sophisticated for primary school children. It can be conjectured to be dependent on the development of two concepts: *conservation* (Piaget 1965), and Resnick's *part-whole schema* (1983). As Ginsburg (1983, 3) suggests, these are closely related concepts and are developing in the primary school years. Although conservation and the part-whole schema are not themselves part of the act of estimating, they can be hypothesized to be necessary prerequisites to decomposition/recomposition.

Conservation

Conservation involves the recognition that numerosity, or quantity, remains constant regardless of changes in position, arrangement, or rotation. To decompose/recompose successfully, as described in the previous section, a student must understand that the number of words remains constant regardless of whether the lines are considered individually or as a group. Conservation concepts must therefore be developed before decomposition/recomposition can be used to determine an estimate. Conservation concepts are not available to young children (Piaget 1960)—they must develop the ability to conserve.

It is a moot point whether interventions are effective to teach conservation concepts. However, it seems highly desirable that young children should be provided with numerous and rich experiences in which they encounter the principle of conservation. Such experiences will presumably enhance the development of the ability to conserve, thereby making estimation problems that require decomposition/recomposition more accessible.

Part-Whole Schema

According to Resnick (1983, 114), "The [part-whole] schema specifies that any quantity (the whole) can be partitioned (into the parts) as long as the combined parts neither exceed nor fall short of the whole." Like conservation, this schema needs to be developed before the decomposition/recomposition process can be used. For example, to estimate the number of cars in a parking lot (the whole), one might first visually partition the lot into sections (parts) and then estimate the number of cars in each section—and thus the number in the entire parking lot. This partitioning requires a recognition that once the whole is separated into parts, the combined parts are equivalent to the whole. Thus, estimation problems that require decomposition/recomposition logically require the development of the part/whole schema by the child.

Many of the informal activities experienced in the primary grades assist in the development of the part-whole schema. Children take apart and reassemble models or find small Cuisenaire rods that together are the same size as a given large rod. Time should certainly be devoted to such activities in the primary grades.

Without conservation and the part-whole schema, children would probably find it difficult to engage in decomposition/recomposition activities. This in turn would hamper their estimation proficiency. Therefore, the development of these abilities should be a focus of attention in the early school years.

NEEDED RESEARCH

The relationships presented in this chapter are, for the most part, logical assertions. Research is needed to test these assertions and to explore their implications for teaching and learning estimation skills. Some suggestions for such research are presented below.

Error Tolerance

The question of the extent to which young children are able to tolerate error lacks research backing. There appears to be nothing in the literature regarding the bounds within which estimates can be made. Can children accept errors that are equally distant from the accurate answer regardless of whether they are of greater or lesser value? If children are introduced to a sequential, noncomputational estimation program in the primary grades, will they develop greater skill with computational estimation than otherwise? Case studies to determine the ability of young children to make estimates and the extent to which they employ guesses or benchmarks when making reasonable estimates would be valuable. Once this observational base is established, it might be possible to develop and determine the influence of instructional treatments designed to help children identify and use guesses or benchmarks when estimating.

Cognitive Structuring

The logical relationships between centering and estimating and between field independence and estimating should be tested. Initially these relationships could be checked by correlation studies or by case studies. If children who are more able to make reasonable nonnumerical estimates are more field-independent and less centering, then treatments that expose children to field independence and noncentering activities might be developed. Experimental research could then be conducted to determine the effect of the treatment on the estimation abilities of young children. The ability to influence the young child's field independence or tendency to center is, of course, still open to considerable debate.

Decomposition/Recomposition

Resnick (1983) proposed the part-whole schema in the context of numerosity. An extension of the model to nonnumerical settings, such as length or time, would be interesting. Also, clinical studies exploring young children's abilities to estimate in these settings would be valuable. Correlation studies focusing on the ability to use the part-whole schema and the ability to conduct estimations would test the relationship suggested in this article. Treatments could be designed to develop for young children the part-whole schema in nonnumerical settings. If these treatments were successful, then they could be tested in experimental studies. Positive results in such studies would further support the link between this psychological variable and the ability to estimate.

SUMMARY

Although much attention is currently being focused on estimation, it has received little notice at the primary-grade levels. Yet it is reasonable to expect that noncomputational estimation, within the limits of young children's experiences and psychological development, would provide a solid base for later estimation skills. It is hoped that the connections between estimation and the psychological variables discussed in this article will provide ideas for teachers, curriculum developers, and researchers as they search for ways to develop estimation ability in the early years.

REFERENCES

Buchanan, Aaron O. *Estimation as an Essential Mathematical Skill* (Professional Paper No. 39). Los Angeles: Southwest Regional Laboratory for Educational Research and Development, 1978.

Burton, Leone. "Mathematical Thinking: The Struggle for Meaning." *Journal for Research in Mathematics Education* 15 (January 1984): 35–49.

Ginsburg, Herbert P., ed.*The Development of Mathematical Thinking.* New York: Academic Press, 1983.

Piaget, Jean, Barbel Inhelder, and Alina Szeminska. *The Child's Conception of Geometry.* Translated by E. A. Lunzer. New York: Harper Torchbooks, 1960.

Piaget, Jean. *The Child's Conception of Number.* New York: Norton Library, 1965.

Resnick, Lauren B. "A Development Theory of Number Understanding." In *The Development of Mathematical Thinking,* edited by Herbert P. Ginsburg, pp. 109–51. New York: Academic Press, 1983.

Reys, Robert E. "Mental Computation and Estimation: Past, Present, and Future." *Elementary School Journal* 84 (May): 547–57.

Reys, Robert E., James F. Rybolt, Barbara J. Bestgen, and J. Wendell Wyatt. "Processes Used by Good Computational Estimators." *Journal for Research in Mathematics Education* 13 (May 1982): 183–201.

Reys, Robert E., Paul Trafton, Barbara Reys, and Judith Zawojewski. "Developing Computational Estimation Materials for Middle Grades." Paper presented at the meeting of the National Council of Supervisors of Mathematics, April 1984, San Francisco, Calif.

Schoen, Harold L., Charles D. Friesen, Joscelyn A. Jarrett, and Tonya D. Urbatsch. "Instruction in Estimating Solutions of Whole Number Computations." *Journal for Research in Mathematics Education* 12 (May 1981): 165–78.

Siegel, Alexander W., Lynn T. Goldsmith, and Camilla R. Madson. "Skill in Estimation Problems of Extent and Numerosity." *Journal for Research in Mathematics Education* 13 (May 1982): 211–32.

Trafton, Paul R. "Estimation and Mental Arithmetic: Important Components of Computation." In *Developing Computational Skills,* 1978 Yearbook of the National Council of Teachers of Mathematics, edited by Marilyn N. Suydamn pp. 196–213. Reston, Va.: The Council, 1978.

Witkin, Herman A., Carol A. Moore, Donald R. Goodenough, and Patricia W. Cox. "Field-dependent and Field-independent Cognitive Styles and Their Educational Implications." *Review of Educational Research* 47 (Winter 1977): 1–64.

7

Developing Estimation Skills in the Primary Grades

Larry P. Leutzinger
Edward C. Rathmell
Tonya D. Urbatsch

A GROUP of primary children was asked to estimate the number of seeds in a pumpkin. After cutting the top off so the children could see the seeds inside, the teacher carried the pumpkin around the room. Each child looked in and made a guess. The guesses of these six-year-olds ranged from one to one million because of their limited experience with number.

When young children try to estimate, predict, or mentally compute, they often suggest or accept answers that adults would consider unreasonable. Their limited perception of the numbers involved and their limited number sense affect their guesses. Young children often think that a number such as 35 is very large and might see little difference between 35 and 1000. To them, 35 and 1000 are both very big numbers.

As early as kindergarten, many opportunities exist or can be created to help children develop an experiential base from which better estimates can be made. As the children are learning estimation skills they are also developing a better sense of number. Teachers can facilitate this development by having children estimate an unknown quantity by (1) comparing it to a known quantity, (2) partitioning it into known quantities, and (3) using mental computation.

A TEACHING STRATEGY FOR ESTIMATION

You can begin estimation activities by having the children make guesses about the quantity of a group of objects or the measure of an object. Questions involving how many, how much, how far, how long, and what fraction can be a starting point. After the children have made their initial estimates, you can give them a hint and let them revise their estimates on this basis.

One type of hint involves comparing the unknown to a known quantity; that is, showing an object or groups of objects similar in size or quantity and stating its numerical value. The children can then compare this known quantity to the unknown one. For example, you can hold up a book (with, say, 156 pages) and ask, "How many pages do you think there are in this book?" After the children have guessed, hold up another book for comparison. "There are 138 pages in this second book. Do you think there are more or fewer pages in the first book? Do you think there are more or fewer than 138 pages in the first book? Do you think there are just a few more pages in the first book or are there lots more? Does the first book have twice as many pages? Do you think the first book has 500 pages? . . . 300 pages? . . . 200 pages?" Through questions or hints you can direct the children's thinking, leading them to make better estimates.

A second type of hint is to partition the unknown into known quantities; that is, show a small part of the unknown and state its numerical value. The children can then estimate the number of such parts in the whole and count or mentally compute to derive a better estimate. For example, hold up a book with 412 pages and ask, "How many pages do you think are in this book?" After the children have guessed, open the book to page 100. "Here are the first 100 pages. Do you think there are just a few more pages or a lot more? There are 100 pages from the beginning of the book to this page. Do you think there are more or less than 100 pages from this page to the end of the book? How many groups of 100 pages are there? Do you think there are 200 pages? . . . 300? . . . 400? . . . 500?"

An activity should not end just because the children have arrived at a reasonable estimate. Strengthen estimation skills by having children verbalize the thinking they used to derive their answers. It helps make them more aware of the procedures they used, and it exposes thinking patterns to other children who may not be using them yet. After the children have made and refined their estimates, you can probe into the thinking behind the guesses. For example, "What were you thinking when you wrote that answer?" "When I showed you the second book with 138 pages, how did you decide how many more pages were in the first book?" "When I showed you 100 pages, how did you use that to help you decide how many pages were in the book?" "Did anyone think about it in a different way?" At first the children will have difficulty putting their thoughts in words, but continued questioning will result in clearer and more frequent responses.

Ask children often to write their estimates. The participation of each child is important. Having them write their answers makes everyone accountable. In addition, after you have given them a hint, they can refer to the written estimate and refine it on the basis of the new information.

Avoid treating estimates as right or wrong, but recognize that some estimates are better than others. Ask questions that indicate boundaries for the range of reasonable estimates. For example, if the known quantity is obviously smaller than the unknown, then estimates should be greater than the number given in the hint. For each estimation activity a range of reasonable guesses should be given. The children should be told why that range was chosen. The size of this range depends on the size of the object being estimated and the hint given. As the children's estimation skills improve, the range for acceptable guesses should become narrower.

The size of the numbers used should be determined by the familiarity children have with those numbers. Typically, numbers to 30 can be used in kindergarten, numbers to 100 in grade 1, numbers to 1000 in grade 2, and numbers greater than 1000 in grades 3 and 4. In the activities that follow, change the numbers to match the abilities of the children.

ACTIVITIES FOR ESTIMATING AND COMPARING TO A KNOWN QUANTITY

Example 1

Do	*Say*
1. Show 17 Unifix cubes linked together. Do not let the students count them.	"How many cubes do you think there are? Write your guess."
2. Now show 14 Unifix cubes linked together. Hold them or place them near the first group for no more than two seconds so the children will not have time to count to find the difference (fig. 7.1).	"Here are 14 cubes. Now how many do you think there are in the first group?"

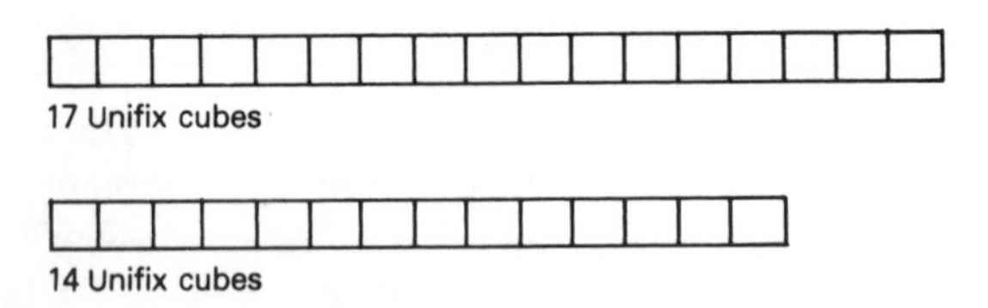

Fig. 7.1

3. Now remove the 14 Unifix cubes from sight. Leave the 17 cubes in view.	"How many cubes were in the second group? " (14) "Are there more or fewer in this

group? " (More)
"Are there more or less than 14?" (More)
"Are there just a few more, or a lot more?" (A few more)
"How many do you think are in this group? . . . Do you want to change your guess? . . . Write down your new guess."

4. Now show the 14 Unifix cubes beside the 17 cubes again.

"Kevin, you wrote 16. How did you decide?"
"Who arrived at their estimate in a different way?"
"What did you think?"

An acceptable range of guesses would be 16 to 18.

Example 2

1. Sort out the dark brown and the green M & M's from a package and place them in a transparent container. Show them to the children (fig. 7.2).

"What do I have in the container?" (Dark brown and green M & M's)
"How many dark brown M & M's do you think there are in the container? Write the number of your estimate."

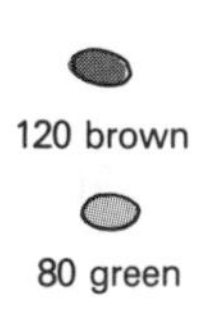

Fig. 7.2

2. Tell how many M & M's there are in the container.

"There are about 200 M & M's in the container. Now how many dark brown M & M's do you think there are? Write down a new estimate if you like."

3. Ask questions to help the children refine their answers.

"Are there more or fewer than 200 dark brown M & M's" (Fewer)
"Is the number of dark brown M & M's close to 200?" (Yes, it's close.)
"Are more than half of the M & M's dark brown?" (Yes)
"Are more than 100 M & M's dark brown?" (Yes)
"Audra, you wrote down 168. How did you think to get that number? Who thought about it in a different way? Would it help to know how many green M & M's there are?" (Yes)
"If there are 50 green M & M's, how many dark brown M & M's do you think there are? . . . That's right, Yolanda, 200 minus 50 is 150, so 150 would be a good estimate."

An acceptable range of guesses would be 120 to 180.

Here are some other estimation activities that involve comparing an unknown to a known quantity.

1. Show the children two stacks of cards. Tell them how many are in one of the stacks. Ask them to estimate the number in the other stack.

2. Have two children of different heights come to the front of the class. Measure the height of one of them. Have the rest of the children estimate the height of the other. Repeat this activity, but use weight instead of height.

3. Show the children two sets of coins. Tell the value of one set. Have them estimate the value of the other.

4. Briefly show a set of coins consisting of quarters, dimes, nickels, and pennies. Tell how many quarters there are. Have the children estimate the value of the set of coins.

5. Have the children close their eyes and raise their hands when they think one minute has elapsed. Repeat, only this time tell the children when a half minute has elapsed.

6. Ask the children to guess the following fractions: What part of the class is left-handed? What part of the class is wearing red? What part of the class is wearing glasses? Is each quantity closer to one-fourth, one-half, or three-fourths?

ACTIVITIES FOR ESTIMATING BY PARTITIONING INTO KNOWN QUANTITIES

Example 1

Do	*Say*
1. Draw a line segment of about 80 cm on the chalkboard. Cut several strips of paper from the bottom of a standard sheet of paper. Tape one strip at one end of the line segment.	"How many strips of paper this length could we tape along the line segment? Write your guess."
2. Help the children find the middle of the line segment. Mark it on the chalkboard.	"Where is the middle of the line segment? How many strips of paper do you think it would take to reach to the middle?" (2)
3. Tape another strip of paper to the board (fig. 7.3).	"Did two strips of paper reach about to the middle of the line segment?" (Yes) "If two strips of paper reach to the middle, how many strips will it take to reach to the end?" (4)

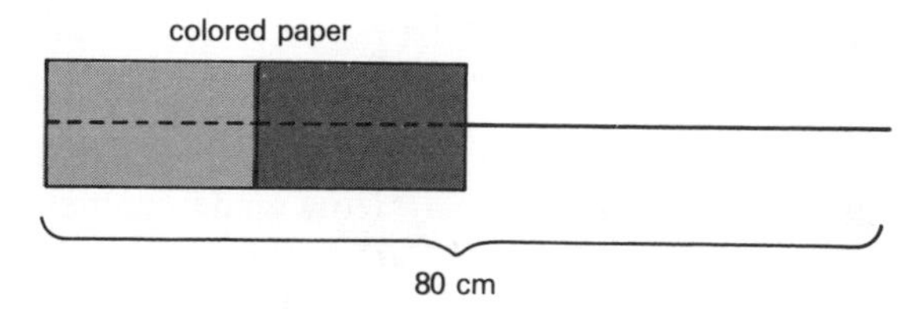

Fig. 7.3

4. Tape two more strips of paper along the line to show that four strips is about as long as the line segment.	"How many strips of paper did it take to reach to the end of the line segment?" (4)

Example 2

1. Make a chain of about 63 paper clips. Show the children.	"How many paper clips do you think there are? Write down your estimate."
2. Show them a chain of 20 paper clips, holding it beside the longer chain (fig. 7.4).	"There are 20 paper clips in this chain. Now how many do you think there are in the long chain? Do you want to change your guess?"

3. Have children imagine putting other chains of 20 paper clips beside the long chain.

"Imagine putting other chains of 20 paper clips beside the long chain. How many would it take to be just about as long?" (3)
"How many paper clips would 3 chains of 20 be?" (60)
"How did you figure that out?" (Added 20 + 20 + 20)
"Who did it in a different way? . . . How did you do it?"

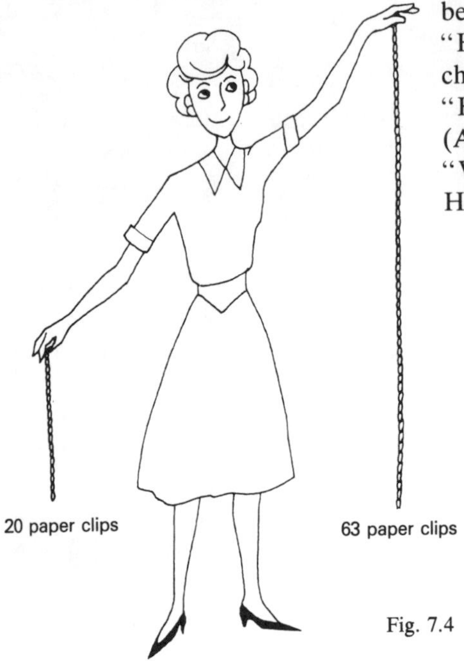

Fig. 7.4

Example 3

1. Show the children a jar of 400 or 500 pennies.

"How many pennies do you think are in this jar? Write down your estimate."

2. Remove 50 pennies from the jar (fig. 7.5).

"I just took 50 pennies from the jar. Now how many pennies do you think there are in all? Remember to think of all the pennies."

3. Remove another 50 pennies from the jar.

"I just took another 50 pennies from the jar. Now how many pennies do you think there are in all? How many pennies have I taken out?" (100)
"Could I take out 100 more pennies?" (Yes)
"How many groups of 100 do you think I could take out? About how

many pennies do you think there are in all? Explain your thinking."

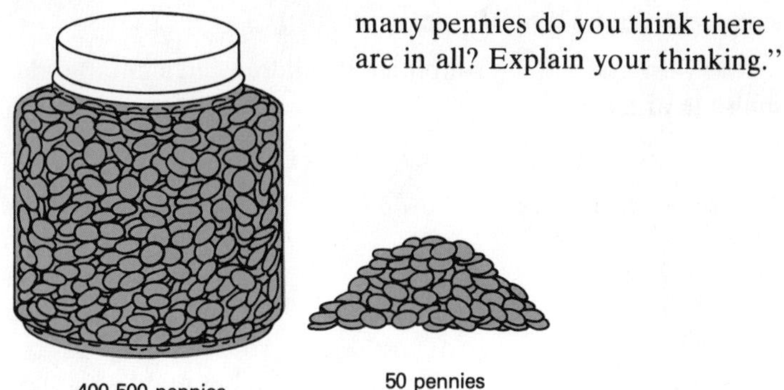

Fig. 7.5

Here are some more estimation activities that involve partitioning unknown into known quantities.

1. Show the children a stack of about 100 sheets of paper. Count out 20 sheets from the stack. Ask them to estimate the number of sheets of paper in all.

2. Ask the children to estimate the length of the teacher's arm in centimeters. Now place a 10-cm rod beside your outstretched arm and have them estimate again.

3. Give the children four note cards to place on their desk edge to edge. Ask them to estimate how many note cards it would take to cover their desk.

4. Show about 30 nickels. Move five aside and count them. Ask how many nickels there are altogether. Ask how much money that is.

ACTIVITIES FOR ESTIMATING BY USING MENTAL COMPUTATION

Mental computation is not a separate estimation process; rather, it is sometimes needed when children estimate by comparing and partitioning with known quantities.

Children need instruction and practice with mental computation before they can use it efficiently for estimation. Following are some activities to show how mental computation can be developed using concrete materials. The emphasis is on verbalizing the step-by-step thinking used while the materials are being manipulated. These activities have to be repeated many times before the children are able to use the mental computations spontaneously. With two or three examples twice a week for a month, children should improve remarkably in their mental computation skill. These examples were selected to show instructional procedures for teaching mental computation. Children need to possess prerequisite skills, such as basic addition and multiplication facts and coin recognition, before attempting the mental computations described.

Adding or Subtracting Multiples of 10

Many estimations are simplified if children can mentally add or subtract a multiple of ten.

Do	*Say*
1. Use bundling sticks to show 34.	"How many tens are there?" (Three) "How much is 3 tens?" (Thirty) "How many ones are there?" (Four) "What number is shown?" (Thirty-four)
2. Show two more bundles of ten (fig. 7.6).	"How many tens are there in this pile?" (Two) "What number is 2 tens?" (Twenty)

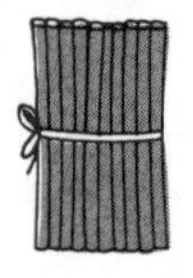
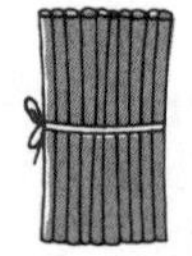
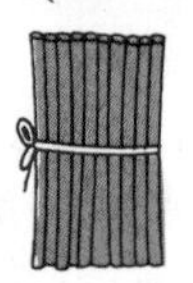
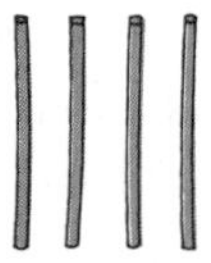

3 bundles of 10 and 4

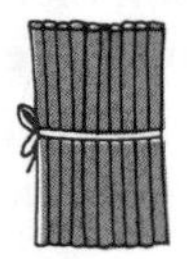
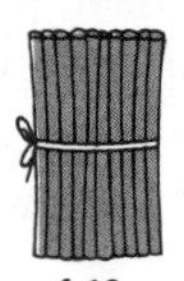

2 bundles of 10

Fig. 7.6

Do	*Say*
3. Move the 2 tens beside the 3 tens. Point to the 3 tens and the 2 tens when you ask how much is 30 plus 20.	"How much is 3 tens and 2 tens?" (5 tens) "How much is 30 plus 20?" (Fifty) "What is 50 plus 4?" (Fifty-four) "To add 34 and 20, think 3 tens plus 2 tens is 5 tens, or fifty. Then add 4 more to get fifty-four."

Adding Single-Digit Numbers

Many mental computations involve adding a single-digit number to a two-digit number. Children usually count on by ones to get the answer; however, they can learn to quickly add the numbers mentally.

Do	*Say*
1. Use 3 dimes and 8 pennies to show 38 cents.	"How many dimes are there?" (Three)

"How much is 3 dimes?" (Thirty cents)
"How many pennies are there?" (Eight)
"How much money is there in all?" (Thirty-eight cents)

2. Show 5 more pennies (fig. 7.7).

"How many pennies are here?" (Five)

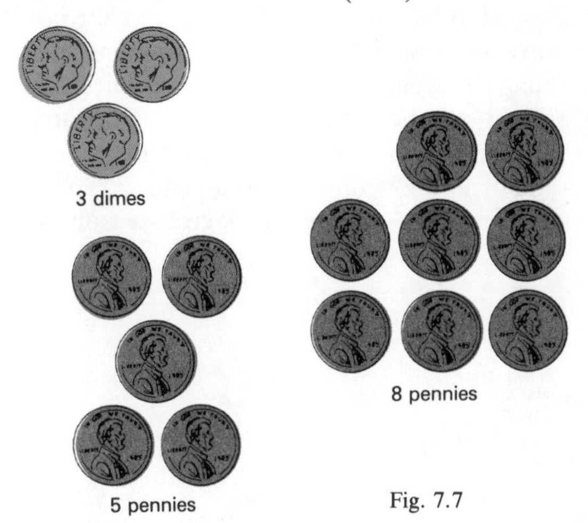

Fig. 7.7

3. Move 2 pennies from the group of 5 to the group of 8 pennies.

"How many more pennies do you need to put with the eight to make ten?" (Two)
"What is thirty-eight plus two?" (Forty)
"How many pennies are left over from the group of five?" (Three)
"What is forty plus three?" (Forty-three)
"Yes, 38 + 5 is 43. We can think, 38 plus 2 more is 40, and 3 more is 43."

Here are some other types of mental computation that can be developed in a similar manner.

1. Mentally counting by fives, tens, twenty-fives, fifties, and hundreds
2. Adding and subtracting multiples of 100
3. Adding and subtracting single-digit numbers and two-digit or three-digit numbers with and without renaming

4. Multiplying single-digit numbers by miltiples of 10 and 100
5. Doubling a number and finding half of a number for multiples of 10 and 100

SUMMARY

Children in the primary grades are capable of learning many estimation skills. However, they need instruction to do so. Also, children's mental computation skills need to be developed if they are to become proficient at estimating. The activities described here are a sampling of some that are appropriate for young children. Children with experiences like these in the primary grades will not only become better at estimating and mental computing but also develop a sense of number. This number sense goes far beyond the ability to compute with paper and pencil. It forms an excellent foundation on which problem-solving and logical-reasoning skills can be based.

8

Critical Balances and Payoffs of an Estimation Program

Chester D. Carlow

MAINTAINING the intuitive freshness and full educational value of genuine estimating over a long period of time is difficult to achieve. It requires a gardener's long-range view of preparing the soil and sowing the seeds. When this is done properly, the harvest may be bountiful.

HISTORICAL BACKGROUND

The development of a systematic estimation program entitled "Concept Development through Estimating and Approximating" was begun in the summer of 1969 at the Ontario Institute for Studies in Education (Carlow 1975–81). The motivation for developing the program grew directly out of general dissatisfaction with mathematics instruction at the time.

A three-member observation team decided to develop the program after visiting classrooms and studying textbooks for a year. The decision was based on the following conclusions:

1. The applications of number system properties in the new math tended to be very formal and cumbersome. By contrast, insights coming out of genuine estimating experiences could be preverbal, simple, and direct.

2. When concrete materials were used in the standard way, the related concepts typically emerged *after* the processes of counting or measuring. Consequently, the concept often got partially lost in the counting or measuring process. In genuine estimating, however, the concepts are strictly up front. The counting and measuring are carried out merely to confirm or refute judgments already made.

3. The occasional approximating being done at the time was typically limited to routine and mechanical rounding-off procedures. The occasional estimating was, for the most part, tied to immediate objectives (as in the

measurement estimates of a particular lesson). Neither of these contributed much to a strong number sense or to a holistic framework for understanding mathematical facts and relationships.

What was needed, it seemed, was the development of two very different sets of skills that we labeled as follows:

1. *Approximating* skills for judging the reasonableness of computational outcomes involving complex numerical expressions. Such judgments are mediated by rounding off, compensating, or other mental operations.

2. *Estimating* skills for making direct, intuitive judgments of quantity in relation to objects in the real world.

The original purpose of the systematic estimation/approximation program was not to do away with the routines and the analyses of existing mathematics programs but rather to build a combined linear/analytic and intuitive/holistic approach that would be superior to either one by itself. The rules and routines would be retained to help make the understandings detailed and precise; the estimating and approximating would be added to support the details with a strong informal background of awareness and understanding.

In the years since 1969 the need for systematic estimating and approximating appears to have increased. In a review of a number of recent research studies, James Hiebert (1984) reports that students still have great difficulty linking form with understanding. After a few years in school, for example, they tend to replace reasonably good problem-solving strategies with shallow, sometimes meaningless, procedures. Also in dealing with fractions many students follow computational rules in a mechanical way, undisturbed by the unreasonableness of answers.

What makes the situation more serious at the present time is the growing accessibility of microcomputers. Without a well-developed sense of mathematical facts and relationships, students have no way to judge the reasonableness of numerical output from a computer.

AWAKENING TO NUMBER AND QUANTITY IN THE WORLD

"Concept Development through Estimating and Approximating" begins in kindergarten with incidental estimating associated with what the child is doing at the moment. Its role is to awaken children to the many quantitative aspects of the world around them. Thus the teacher "wonders" how many blocks can be stacked before they fall over, or whether there are enough pencils to go around.

Also there are estimates relating number to the everyday world through linear measure, area, and volume. For example, children estimate how many

of them it will take to stretch the length of the classroom, how many handprints it will take to cover the paper, or how many children can squeeze into the big box.

The first estimates are critical for avoiding the pitfalls of wild guessing, partial counting, mechanical routines, unrealism, and oversimplicity. The estimating tasks must be easy enough for children to *want* to make direct judgments of quantity without the mediating procedures (such as counting) that destroy the intuitive grasp of the whole situation. Yet they must be challenging enough to involve real and significant judgments.

Incidental estimating plays an important role through the grades by enhancing the understanding of topics currently being studied. However, planned estimation lessons (of about two or three estimates each) are required to provide systematic preparation for things to come.

PROVIDING AN INTUITIVE PREVIEW OF THINGS TO COME

One potential payoff of estimation lessons is the often unconscious, partial learning of topics before they are studied formally. Thus, in "Concept Development through Estimating and Approximating" there are preaddition lessons, premultiplication lessons, and lessons that hint at many other mathematical facts and relationships. From the student's point of view, the focus of these lessons is on making good estimates. From the teacher's point of view, the objective is two-fold:

1. To help the students become good estimators
2. To help them become familiar with, or subconsciously aware of, *some* of the facts and relationships reflected in the materials

The *some* is stressed to point out the freedom that children must be given to develop intuition in their own individual way. The learning materials may provide enriched environments, and the teacher may encourage light-touch discussions; however, any heavy-handed emphasis on tight, logical sequencing or immediate measurable outcomes, which works well in other mathematics learning, will tend to retard the development of intuition.

Our lessons on direct estimating of number illustrate the attempt to develop children's estimation ability and at the same time give them an intuitive preview of things to come. In order to make direct estimates, children must internalize appropriate standard referents or perceptual anchors. A perceptual anchor is a quantity whose size is readily perceived in different settings; thus, a child can use internalized perceptual anchors to make comparative estimates of other quantities. Children, on their own, normally internalize no more than two perceptual anchors, their own weights and heights, and both of these have limited applicability to estimating number.

Therefore, if one wants to promote direct estimates of number, one must develop the appropriate perceptual anchors. For the preaddition and premultiplication estimating we developed just three perceptual anchors—*anchor 10, anchor 20,* and *anchor 100.* We use each anchor for a broad range of estimates, roughly ranging by a factor of 2.5 below and above the anchor number. Thus, anchor 20 is used as a referent for estimates of sets with as few as eight and as many as fifty elements. To avoid the heavy-handedness mentioned above, we softened the focus on computation by hinting at other things as well. A few examples (fig. 8.1) will illustrate the general nature of these estimates. These anchor-20 estimates, while useful in developing the ability to estimate, can also promote conservation of number, hint at some addition and multiplication facts coming up later (such as 5×5 and 3×4), and hint at the relationship between areas of rectangles and areas of parallelograms.

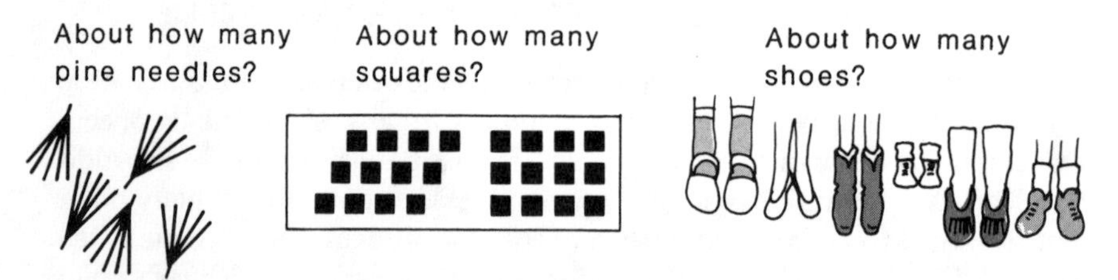

About how many pine needles?
About how many squares?
About how many shoes?

Fig. 8.1

The anchor-100 estimates in figure 8.2 include randomly arranged objects to help prevent excessive dependence on patterns, other sets of objects to suggest the 7×7 multiplication fact, and the idea that the area of a circle is a little more than three-fourths the area of a square having the same diameter.

Other kinds of estimates of number in the program do not involve perceptual anchors. Like the incidental estimates, these estimates use as referents a part of all the objects present at the time (fig. 8.3). Such estimates are

About how many objects in each?

Fig. 8.2

included not only to develop estimation ability but also to extend the young student's general awareness of number beyond the normal range.

About what fractional part of the stars is enclosed?

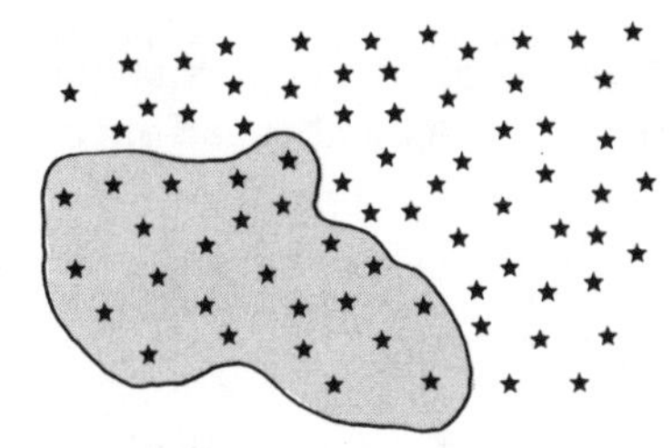

About what percent of the square is shaded?

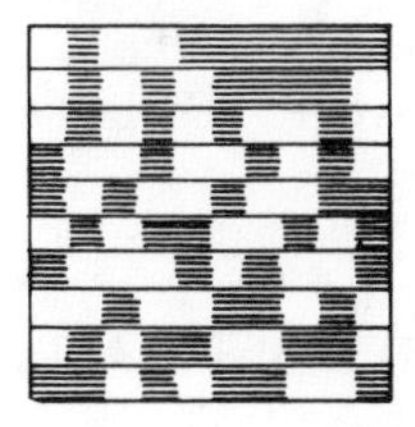

About how many hairs in the comb?

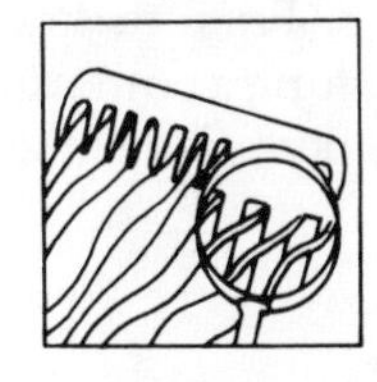

Fig. 8.3

CONFIRMING AND CONSOLIDATING AFTER FORMAL LEARNING HAS TAKEN PLACE

Many of our estimation lessons in grades 4, 5, and 6 occur after the topics have been dealt with formally in the regular mathematics program. The role of these estimates, therefore, is to confirm and consolidate rather than to prepare the way for learning. In addition, two chapters in the book for grades 5 and 6 are devoted to the development of approximation skills—skills for judging the reasonableness of computational outcomes. A few examples will illustrate these lessons:

1. $278 \times 47 \approx$ (*a*) 300×50
   (*b*) 200×50
   (*c*) 300×40
2. $486 \div 54 \approx \square \div \square$
3. $24 \times 44 \times 77 \approx \square \times \square \times \square$

Students learning the approximation skills will have had routine rounding-off experiences and will now proceed to more sophisticated and holistic compensating techniques in order to achieve closer approximations.

The best alternative in the first exercise listed above is

$$278 \times 47 \approx 300 \times 40.$$

Even though the 40 is a little too low, students will learn that if the second digit of 278 is *increased* by 3, the second digit of 47 must be *reduced* by 4 or 5 to compensate (because 47 is greater than 27).

In the second exercise, 450 and 50 are good approximating numbers:

$$486 \div 54 \approx 450 \div 50$$

In division exercises (as students should learn from working with equivalent fractions), changes in the dividend are compensated by changes *in the same direction* in the divisor.

In the third exercise,

$$24 \times 44 \times 77 \approx 20 \times 50 \times 80.$$

Increasing 44 to 50 is not enough to compensate completely for reducing 24 to 20. Thus 77 is also increased to 80. And though experienced approximators may feel it is still not quite enough, they let it stand as a fairly good approximation.

From these examples it can be seen how estimating and approximating form a powerful neans of enriching the understanding of number and operations on numbers.

CRITICAL BALANCES IN ESTIMATING

It is well to be somewhat tentative about proclaiming critical balances in the teaching of anything as recently developed as systematic estimating. What may be required in one strategy may be harmful in another. Nonetheless, real estimating simply cannot be taught in the tightly sequenced and rule-bound approach commonly used in other mathematics learning without destroying both the integrity and potential educational value of the instruction. Thus, there seem to be a number of critical balances or "golden means" that must be maintained for the successful teaching of estimation. If one strays very far from these, the results can be devastating. Following are nine such critical balances.

1. *Develop the "right" number of perceptual anchors—about three in each of five or six areas.* The need for perceptual anchors arises because estimates, to be educationally valuable, must be almost effortless and holistic. Then students have the freedom to notice facts, relationships, and qualities embedded in the things being estimated. With too few anchors, the domain of genuine estimating—and with it the number of relationships that can be addressed—is restricted. With too many anchors, the learning task becomes too complex and confusing.

After much deliberation we selected the following "lean" system of anchors, including the anchors for numbers already mentioned.

Anchors for Estimating in Various Domains

Number	Length/ distance	Area	Volume	Mass	Angle	Fractional numbers
10	1 m	100 cm²	1 ℓ	1 kg	90°	halves, thirds, and fourths
20	30 cm	1000 cm²	30 ℓ		30°	
100	30 m	10 m²	1 m³			percents (%)
large numbers	100 m		30 m³			per mil (‰)
	small measures					

2. *Show the objects to be estimated for the right amount of time—about one to three seconds for each showing, depending on the complexity of the things being estimated.* Too brief a showing leads the children to believe that the teacher doesn't really expect a careful estimate. Prolonged showings promote habits of analyzing and partial counting that destroy the direct intuitive grasp of genuine estimating. In the words of William Wordsworth:

> Our meddling intellect
> Mis-shapes the beauteous forms of things:—
> We murder to dissect.
>
> —"The Tables Turned"

3. *Have the right number of showings for each estimate—exactly two.* When estimates are made in connection with large objects in the environment, the showings are achieved by having the children look, then look away.

4. *Have the right proportion of random and patterned arrangements—about half and half.*

5. *Have the right number of estimation lessons each week—about two or three.* In addition to estimation lessons, some estimating directly related to, and carried out during, the regular mathematics class is essential from time to time. However, this must be done in moderation. For example, to encourage students to make estimates in connection with all or most measurement activities is disastrous. Its perceived uselessness quickly undermines children's attitudes about all estimates.

6. *Have the right number of estimates for each lesson—about three.*

7. *Have the right amount of follow-up discussion for each lesson—a "light touch" discussion of five to ten minutes.*

8. *Have the right amount of student checking.* Our approach to checking follows:

a) All children count in unison to check very easy estimates related to anchor-10 and -20 estimates.

b) Two students count or measure and report immediately for other easy-to-check estimates.

c) Two pairs of students check and report later for more difficult checking requiring more than five minutes. (If pairs differ, they must check again.)

d) Students as a group are not asked to estimate and then check their own estimates because it quickly leads to the perception that estimating is not important.

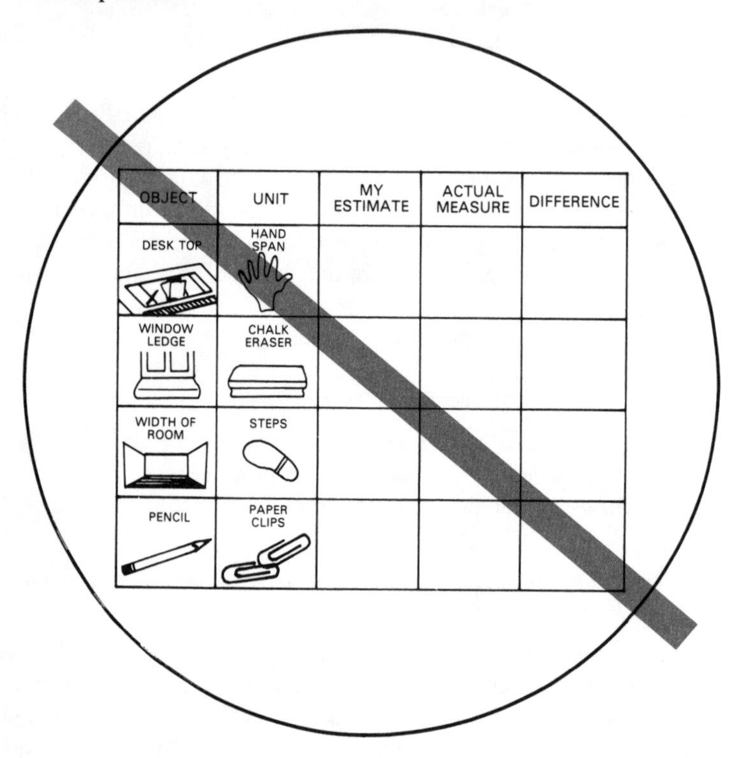

OBJECT	UNIT	MY ESTIMATE	ACTUAL MEASURE	DIFFERENCE
DESK TOP	HAND SPAN			
WINDOW LEDGE	CHALK ERASER			
WIDTH OF ROOM	STEPS			
PENCIL	PAPER CLIPS			

9. *Have the right amount of training for teachers to develop their own estimating skills—a minimum of one three-estimate lesson each month in conjunction with regularly scheduled meetings until all teachers become fairly good estimators.*

Failure to maintain these nine balances tends to result in student frustration, loss of interest, inability to make good estimates, partial counting, limiting the relationship between estimating and other mathematical content, estimating on the basis of extraneous cues, failure to capitalize on purposeful measurement activities, and failure of teachers to keep up with their students. Each of these results seriously undermines the genuineness and potential value of the estimates.

A few comments on critical balance number 4 will indicate the kinds of problems that arise when a critical balance is not maintained. As we argued earlier, under the right conditions estimating can be an obvious and powerful vehicle for helping children develop the ability to conserve number. In fact, the very notion of conceptual anchor implies that children have learned to recognize the quantity represented by the anchor in a wide variety of contexts. It follows, therefore, that if critical balance 4 is not maintained and sets of objects being estimated contain too many patterns or regularities, then the generality of the anchors will not be achieved, conservation will not be promoted, and the genuineness of the estimates themselves will be undermined. However, in an already crowded curriculum systematic estimating can best be justified if it can be shown to prepare the way for, enhance, and consolidate learning more effectively than an equal amount of time spent in some other way. And the only way this can occur is to present some patterns and regularities among some of the things being estimated, since in this way one may hint at, and gradually develop, an appreciation of a great variety of mathematical facts and relationships.

The balances for developing sophisticated approximation skills are less documented at this time but nevertheless appear to be just as critical. One must allow enough time for the unhurried exploration of the relationships between the numbers in an expression yet not enough to encourage partial calculations. One must sequence so that the approximations are easy enough to promote a reasonable level of success but difficult enough to involve significant judgments. One must give enough guidance to promote satisfactory progress but not so much that the approximating strategies become rigid and routine.

Although other critical balances may be found in the future, these nine are enough to indicate that systematic estimating and approximating, like genuine problem solving, may be among the most difficult to teach of all the goals of mathematics. The approach to teaching them is much different from that of teaching computational skills. And the payoffs, as we shall see, are also very different.

POTENTIAL PAYOFFS

The greatest potential payoff of an extensive estimating/approximating program is the greatly enriched preparation for meaningful learning. It is well known that the quality of learning depends on the existing perceptions, intuitions, and understandings that the student brings to the learning situation. As expressed by the cognitive psychologist David Ausubel (1968, vi),

> If I had to reduce all educational psychology to just one principle, I would say this: The most important single factor influencing learning is what the learner already knows.

What the learner knows through estimation and approximation, properly taught, tends to be intuitive, nonverbal, relational, and real. What the learner knows through typical classroom activities supported by typical mathematics textbooks is more formal, verbal, factual, and (in many children's eyes) arbitrary.

A second potential payoff for an extensive estimation program comes from a greatly increased familiarity with the quantitative aspects of the environment. This helps to demystify mathematics for many students and makes mathematics more relevant for all.

A third potential payoff has to do with increased motivation. Two different factors may be involved. From the cognitive side a large number of children do not do well in mathematics when it is taught primarily in terms of rules, linear progressions, and verbalizations that depend heavily on good memory and good self-monitoring skills. Some of these children may be more successful, and hence better motivated, by an alternative mode of understanding—a more intuitive, informal, and holistic one. In addition, the estimation lessons can be so different and so enjoyable that motivation may be increased.

A fourth payoff is the substantial saving of time that may result. For example, even rudimentary approximating skills would prevent the child from getting lost in time-consuming formalities such as the following:

$$\begin{array}{rrrr} & & 9 & \\ & 0 & \not{1}\not{0} & 10 \\ & \not{1} & \not{0} & \not{0} \\ - & & 9 & 9 \\ \hline & 0 & 0 & 1 \end{array}$$

A fifth potential payoff is a general improvement in teaching. When teachers harvest good estimations and approximations as they prepare the ground and sow the seeds for further, more holistic understanding, the experience may generalize to other situations. A substantial enrichment of the present narrow, flowchart approach to teaching could result.

In summary, systematic estimating and approximating skills over several domains can be taught effectively only if curriculum developers and teachers maintain a number of critical balances. If this is well done, the payoffs can be considerable indeed.

REFERENCES

Ausubel, David P. *Educational Psychology: A Cognitive View.* New York: Holt, Rinehart & Winston, 1968.

Carlow, Chester D. *Concept Development through Estimating (and Approximating).* Limited edition for classroom trials. Toronto: Ontario Institute for Studies in Education Press, 1975–81.

Hiebert, James. "Children's Mathematics Learning: The Struggle to Link Form and Understanding." *Elementary School Journal* 84 (May 1984): 497–513.

9

Estimation and Children's Concept of Rational Number Size

Merlyn J. Behr
Thomas R. Post
Ipke Wachsmuth

WHAT is it that children need to know about fractions in order to estimate them? What misunderstandings are represented, for example, by the fact that only 24 percent of the nation's thirteen-year-olds were able to correctly "estimate" 12/13 + 7/8 by selecting the correct answer from among 1, 2, 19, 21 (Post 1981)? The popular choices of 19 and 21 (chosen by 28 percent and 27 percent, respectively) certainly suggest that many children do not understand that 12/13 and 7/8 are each close in size to 1. In our view, this example suggests that success in estimating 12/13 + 7/8 depends on an understanding of the *size* of the respective rational numbers. We take the position in this article that an understanding of the size of numbers—whole numbers, fractions, decimals—is essential to the ability to find estimates for them. We also believe that estimation can be used to develop an understanding of number size.

In a general sense, in order for a thing to have the attribute of size, it must be conceptualized as a single entity. To speak of the size of a pile of pennies, for example, suggests that the pile is conceived as one entity. Do children understand a fraction as a single entity or as two separate numbers? A fraction symbol is composed of two whole-number symbols—the numerator and the denominator. For a child to understand a rational number as one entity, the ability to coordinate the meaning of these symbols is required. To determine the size of a rational number requires an understanding of the *relationship between* the numerator and denominator. The numerator, the denominator, and the relationship between them must all three be coordinated in order to estimate correctly or understand the size of a given rational number. Many children do not coordinate these three ideas but handle numerators and denominators separately. This is strongly suggested by the fact, noted earlier, that 55 percent of thirteen-year-olds selected either 19 or 21 (the sum of the numerators and denominators, respectively) as an estimate for 12/13 + 7/8.

This article will discuss in detail two rational number estimation tasks and clarify the thinking strategies used by children as they complete them. The article will also suggest other activities that will promote children's ability to estimate and understand the size of rational numbers.

TWO TASKS FROM RESEARCH

The information that we present here about children's strategies in dealing with rational numbers emanates from experimental work of the Rational Number Project (see Behr, Wachsmuth, Post, and Lesh [1984] for details). Fourth- and fifth-grade children exhibited these strategies during thirty weeks of project teaching experiments. The instruction emphasized the use of manipulative aids and considered five topics: naming fractions, identifying and generating fractions, comparing fractions, adding fractions, and multiplying fractions. Each child was individually interviewed on eleven separate occasions. The interviews were conducted approximately every eight days during the thirty-week instructional period. Each interview was audio- or video-recorded and later transcribed. The two tasks from the project that will be discussed in this article are concerned with comparing fractions and adding fractions, respectively.

Order and Equivalence

The estimation of a fraction has to do with the notion of the closeness of one number to another. Related to this notion is the question of whether two fractions are equal, and if not, which one is less.

During interviews we asked children to order (decide which is greater) fractions of three basic types: same-numerator fractions, same-denominator fractions, and fractions with different numerators and denominators. Our analysis suggested that five or six different strategies were used by children for each of the three types of conditions. The majority of these were valid and in some way recognized the relative contributions of both numerator and denominator to the overall size of the fraction. In some situations, however, children focused only on the numerator or only on the denominator and as a result made incorrect conclusions. In other instances they compared each to a common third number (usually 1/2 or 1) and were successful in ordering the given fractions. For example, "2/5 is less than 5/8 because 2/5 is less than 1/2 and 5/8 is greater than 1/2."

Even after extensive instruction, some children were at times negatively influenced by their knowledge of whole-number arithmetic and as a result reached inappropriate conclusions. For example, "1/17 is less than 1/20 because 17 is less than 20." We called this strategy *whole-number dominance*.

The least successful students on these tasks were those that had the most difficulty coordinating numerators and denominators. Apparently they had difficulty with these multiple relationships or were unable to remember them long enough to make a decision about the original question. The nature of this difficulty becomes apparent when one attempts to order 5/8 and 3/5 without using some type of abstract algorithm or reference to a manipulative aid to reduce reliance on memory alone. (Try it and see!) We also found that children's ability to deal with order and equivalence declines further when "difficult" fractions are used and when the ordering task is embedded in a verbal problem-solving situation.

It seems reasonable to assert that it is not possible to estimate satisfactorily the sum, product, difference, or quotient of two rational numbers unless one has the ability to determine the relative size of two or more rational numbers. It is essential, therefore, that students understand this basic concept of relative size before they can be expected to estimate rational numbers accurately and with an acceptable degree of understanding. As the results from this task show, many students do not have a good understanding of relative size and often use incorrect strategies.

Construct-a-Sum

The task *construct-a-sum* was given to sixteen children midway through grade 5. These children were subjects in teaching experiments of the Rational Number Project during their grade 4 and grade 5 years (Behr, Wachsmuth, and Post 1985).

The first of two versions of this task consisted of numeral cards on which the whole numbers 1, 3, 4, 5, 6, 7 were written and a form board as shown in figure 9.1.

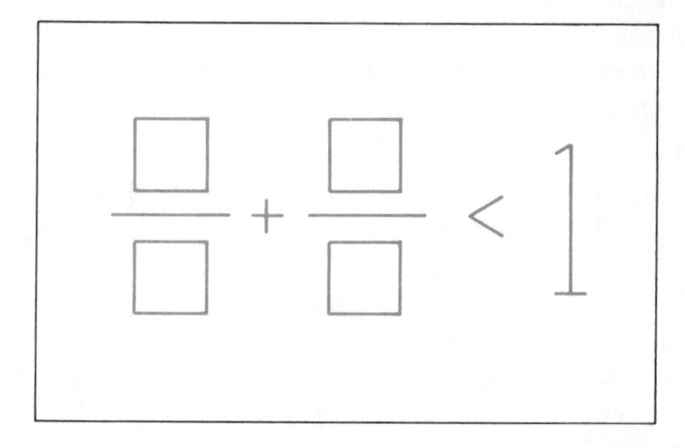

Fig. 9.1

The second version used the same form board but numeral cards with 11, 3, 4, 5, 6, 7. Version 1 was presented about five-sixths of the way into the children's grade 5 year; the second version was presented shortly thereafter. In each instance, the children were directed to place numeral cards inside the boxes to make two fractions that when added are as close to 1 as possible but not equal to 1. To discourage the use of computational algorithms, subjects were encouraged to estimate, and a time limit of one minute was imposed. After completing the task, subjects were asked to "tell me how you thought when solving this problem."

From children's explanations, responses were partitioned into five categories plus an "other category." These categories suggest cognitive strategies that children used to perform the task. Especially observable in the responses is the variation in the subjects' use of estimation, fraction order and equivalence concepts, and reliance on a correct or incorrect computational algorithm. The various strategies are described below.

Correct reference point comparison. Responses in this category reflect a successful attempt to estimate the constructed rational number sum by using 1/2, 1, or some other self-constructed fraction as a point of reference. The spontaneous use of fraction equivalence and rational number order is evident. The following interview excerpt is illustrative:

Bert: [Using 1, 3, 4, 5, 6, 7, he makes 3/6 + □/□, pauses, thinks, changes to 4/6 + □/□ and finally to 5/6 + 1/7.]

Interviewer: Tell me how you thought about the problem.

Bert: Well, umm . . . , 5/6 is . . . well, a sixth is larger than the seventh, and so there [i.e., 5/6] is one piece away from the unit covered, a seventh is smaller, so a seventh can fit in there [i.e., between 5/6 and the whole] without covering the whole.

Mental algorithmic computation. Explanations of student responses indicate here that the subject used mental computation to carry out a correct standard algorithm (e.g., common denominator) to determine the actual sum of the generated fractions. The spontaneous use of fraction equivalence and rational number order is also evident in this category of responses.

Kristy: [Using 1, 3, 4, 5, 6, 7, she makes 1/3 to □/4, then changes to 1/3 + 4/5.] If you find the common denominator, twelve. But [referring to 1/3 + □/4] . . . , and then four times one would be four [changing 1/3 to 4/12]. But then [explaining the change of □/4 to 4/5] three times . . . I didn't have a two or anything [among the number cards given and remaining], and I used up my three, so. . . .

Observe what Kristy is apparently doing: 1/3 is equivalent to 4/12. How many more twelfths are necessary to get close to one? This is determined from □/4 or 3 × □/12; so realizing that she has only 5, 6, or 7 to choose for the box, which in turn give 15/12, 18/12, and 21/12—too many twelfths—she changes □/4 to □/5 and now must do the same type of thinking with fifteenths.

Incorrect reference point comparison. Students' explanations of their responses in this category indicate that they attempted to estimate the constructed rational number sum by using 1/2, 1, or some other self-constructed fraction as a point of reference. Little understanding of fraction equivalence and rational number order was evident.

Mental algorithmic computation based on an incorrect algorithm. Responses indicate that the subject used mental computation based on an incorrect algorithm to compute the actual sum.

Ted: [From 1, 3, 4, 5, 6, 7, he made 5/6 + 4/7.] Well, first I thought, I tried to figure out what would come closest to 1, and I found out that 5/6 and 4/7 would come the closest . . . 'cause I used the top number . . . (if I added them) 9/13.

Gross estimate. Explanations here suggest that the subject made a gross estimate of each rational number addend but did not make a comparison to a standard reference point and did not use fraction equivalence or rational number ordering.

Ted: [From 11, 3, 4, 5, 6, 7, he makes 3/11 and 4/7.] I wanted to use up the little pieces from the top . . . then use the highest number of pieces from the bottom. . . .

We were somewhat surprised to find a large percentage of the responses given by these children late in grade 5 to be incorrect. A large percentage of the responses (20 of 41, or 49%) suggested a processing of fraction addition that showed little understanding of fraction size. At times they suggested mental application of an incorrect algorithm, such as adding numerators and denominators. Children who gave these responses were in the middle or low range of mathematical ability. Correct responses were given mostly by children with average and above-average ability in mathematics.

OTHER TASKS AND TEACHING STRATEGIES

We believe that it is important to help children develop a concept of the size of rational numbers. This requires that they be able to view fractions as single entities and as numbers in and of themselves. When this concept has been internalized, it will then be possible to help children develop an ability to estimate the size of rational numbers. As skill is developed in estimating rational numbers, it feeds back to improve a child's concept of rational number size. Thus, the concept of rational number and skill in estimation can be developed in such a way that they go hand in hand and facilitate each other.

Practice on the tasks we present and on similar tasks will improve children's ability to do the construct-a-sum task and to estimate the result of operations on numbers; for example, to estimate 12/13 + 7/8. We present the first of these in the context of a possible dialogue between teacher and class. Although the dialogue is somewhat "idealized," it does demonstrate situations that are likely to surface in discussions with children.

Activity 1: How close can you get to 1?

Objective: To help children observe how changing numerators or denominators or numerator-denominator pairs causes a change in the size of the fraction.

Teacher: Can you name a fraction that is close to 1?

S1: Three-fifths.

Teacher: That's close to 1; how close is it?

S2: It's 2/5 close.

Teacher: Well, then, can you give a fraction that is closer to 1?

S2: I think 4/5 would be closer. Because 4/5 is *only* 1/5 away.

This response required only a numerator change. Sometimes it might be necessary for the teacher to give the first fraction in order for certain observations to be possible.

Teacher: Is it possible to get even closer to 1?

S3: It's harder now, but I think 5/6 is only one away from 1 [i.e., 1/6 away from 1], and 1/6 is less than 1/5, so 5/6 should be closer to 1 than 4/5.

(Ten-year-old Kathy suggested this approach.) As children gain skill, these answers are not atypical. Observe the number of things that must be remembered to give a response like Kathy's!

1. 4/5 is 1/5 away from 1 whole.
2. To reduce the difference, find a fraction that is less than 1/5.
3. This can be done by increasing the denominator.
4. 1/6 will do—1/6 is less than 1/5.
5. 5/6 is 1/6 away from 1 whole.
6. Therefore, 5/6 is closer to 1 than 4/5.

Kathy's solution is an impressive mental feat when viewed from this perspective. From this activity, and practice on many more items, the child gains familiarity with the relation between fraction size and changes in the numerator, the denominator, or both. An especially important observation is that when the numerator increases and the denominator decreases, the fraction increases rapidly. Likewise, the fraction decreases in size when the numerator decreases and the denominator increases.

Children pick up the notion of "one part away from" (that is, one unit fraction away from) rather quickly. With practice it can be a powerful means of getting fractions close to (and closer to) a given number. The task can be varied with challenges such as these: (*a*) Name a fraction *very close* to 1; (*b*) name a fraction so close to 1 that you think no one could give one closer (this second question leads to the observation that you can keep giving fractions that are increasing in size yet never get a fraction as great as 1); and (*c*) name a fraction so close to 0 that you think no one could give one closer (this leads to the observation that there are smaller and smaller positive rational numbers). Some rather profound mathematics comes up in a natural way. Other variations are possible by changing the reference number 1 to other "usable," "comfortable," and "handleable" numbers, such as 1/2, 1/3, 3/4, 2/3, 2, and 2 1/2.

Variations on the Theme

Activity 2: Get close, and closer, to 1, staying above 1.

Objective: Same as activity 1.

Activity 3: Get close, and closer, to 1, while alternating above and below 1.

Objective: Same as activity 1.

Another series of related estimation activities requires children to produce or approximate a target number by manipulating some set of rational numbers. For example:

Activity 4: Given set $A = \{1/2, 1/3, 1/4, 1/5, 1/6, 1/7, \ldots\}$, get close (and closer) to the target number (3/4, for example), using three numbers from set *A* and the operation of addition (or any combination of operations; or four numbers; or five numbers, etc.). Obviously the makeup of set *A*, the target number, and the task conditions can be manipulated to serve related goals.

Activity 5: From the set $B = \{1/2, 1/3, 2/3, 1/4, 2/4, 3/4, 1/5, 2/5, 3/5, 4/5\}$, try to construct given target numbers (1 1/12, for example). Use as few (or as many) numbers from set *B* as you can. Numbers can (or cannot) be used more than once.

Activity 6: A variation on activity 5: This activity is for two players. Each selects two fractions or whole numbers from the board below. They then add, subtract, multiply, or divide those two fractions. The score is determined by locating the interval on the number line within which the new fraction resides and awarding points as indicated. Four turns by each player constitutes a game. Three points are awarded if the numbers 0, 1/2, 1, 1 1/2, or 2 are "hit" exactly.

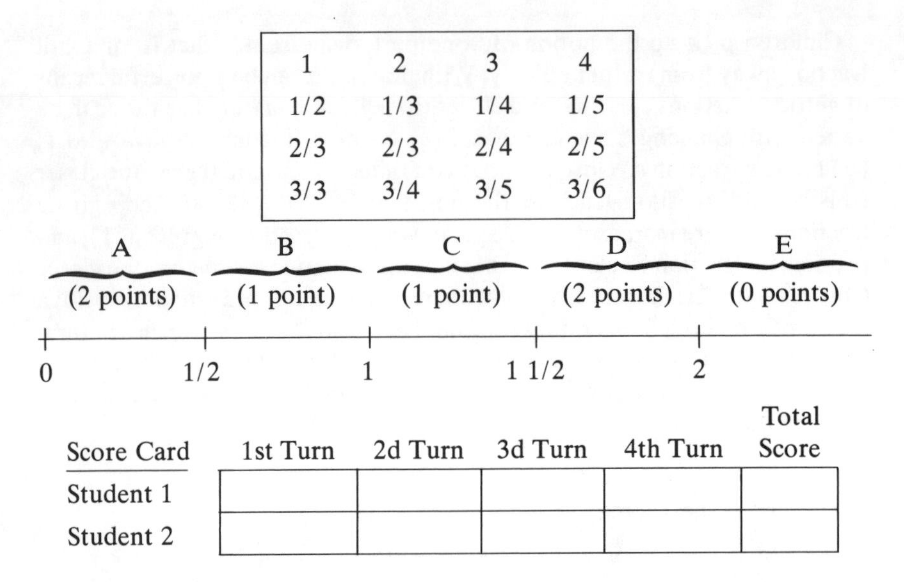

Score Card	1st Turn	2d Turn	3d Turn	4th Turn	Total Score
Student 1					
Student 2					

The entries on the fraction board, the dimensions of the number line, and the point allocations can be manipulated to suit the objectives under consideration. When a constraint is placed on the amount of time that a child can have to produce an answer, an estimation of the size of the sum is required. The estimation of a sum depends on good estimates of the addends.

Activity 7: Fraction estimation with a calculator. Several calculators are now available with a fraction mode. The Casio fx350 Scientific Calculator is one of these. At this writing, it is available for less than $15. The fx350 uses a small bracket to denote the fraction slash or bar. Thus 1⌋2 is interpreted as 1/2, and 1⌋2⌋5 means 1 and 2/5. Fraction-to-decimal conversions are available with a single additional depression of the fraction key. If you have access to this or a similar calculator, your students will benefit from the following activity. Once again the student is asked to hit the target.

Step 1: Student A enters a number in the calculator. This can be an integer, a fraction, or a mixed number (you may wish to place limits on this number initially). Student A then identifies a target area or region, for example, between 20 or 30; or you may wish to do this at first. Student A then passes the calculator to student B.

Step 2: Student B must now push the add, subtract, multiply, or divide key and enter any number in fraction form, for example, 1/2, 4/2, 13/5, and so forth. Student B will then push the = key. If the number appearing on the display is within the target area, student B earns one point. If not, the calculator is passed back to student A *as is,* and student A must attempt to reach the target area using the same procedure. This continues until the target is reached. Here is a sample game:

1. Student A enters "24" and identifies a target area of 10 to 11. Passes calculator to student B.
2. Student B presses "÷" and "1/2" (1⌋2), presses "=," and gets "48." Passes calculator to student A.
3. Student A enters "×" and "1/4" (1⌋4), presses "=," and gets "12." Passes calculator to student B.
4. Student B enters "−" and "6/5" (6⌋5), gets "10⌋4⌋5" (10 4/5), and wins the game. Student B gets one point.
5. Repeat with new numbers.

The authors have used this latter activity with in-service elementary teachers and have found the game to be "nontrivial." Even mathematically sophisticated professionals can gain important insights into estimation processes and rational number understandings by participating in these deceptively simple activities.

CONCLUSION

Many of the activities that were suggested here to develop estimation skills with rational numbers are similar in format to those that would have been appropriate in the development of whole-number estimation skills. This is an economical situation in that it enables you to easily adapt estimation activities that are already in your repertoire. Over the years, NCTM publications have provided many other such suggestions, and many more are included in other articles in this yearbook.

Our research indicates that the learning of rational number concepts is more complex than we originally thought and is in fact very difficult for many children. One of the main areas of need is to help children think about fractions as numbers in their own right and not only as parts of a whole, as ratios, or as the quotient of two integers. The activities suggested here all emphasize the concept of rational number size, stressing that fractions are themselves single numbers. Estimation is an excellent vehicle for developing this concept. In fact, we have seen the mutually supportive roles played by estimation with rational numbers and the concept of rational number size.

REFERENCES

Behr, Merlyn J., Ipke Wachsmuth, and Thomas R. Post. "Construct a Sum: A Measure of Children's Understanding of Fraction Size." *Journal for Research in Mathematics Education* 16 (March 1985):120–31.

Behr, Merlyn J., Ipke Wachsmuth, Thomas R. Post, and Richard Lesh. "Order and Equivalence of Rational Numbers: A Clinical Teaching Experiment." *Journal for Research in Mathematics Education* 15 (November 1984):323–41.

Post, Thomas R. "Fractions: Results and Implications from National Assessment." *Arithmetic Teacher* 28 (May 1981):26–31.

10

Estimating Fractions: A Picture Is Worth a Thousand Words

Gary E. Woodcock

ESTIMATION, like many other aspects of mathematics, needs to be approached not only concretely but also visually. Diagrams play an important role in the development of estimation skill and also lead to a better understanding of computational rules and problem solving. This article describes the approach I use for developing estimation skills for the multiplication of fractions.

A sequence of activities is recommended. For each, students draw pictures for a set of exercises. Then the teacher asks them questions that encourage them to see patterns and make generalizations. I believe it is important for students to form and state their own conclusions, since it causes them to be actively involved in, and accept responsibility for, their learning.

Basic to estimation work with fractions is the ability to order them. Activity sets A through D involve ordering and estimating with fractions less than1, and sets E and F do the same with fractions greater than 1. Sets G and H are a synthesis of the previous sets.

Set A: Draw the four rectangles described below.

1. Draw a rectangle that is about 1/10 the size of the one shown in figure 10.1.

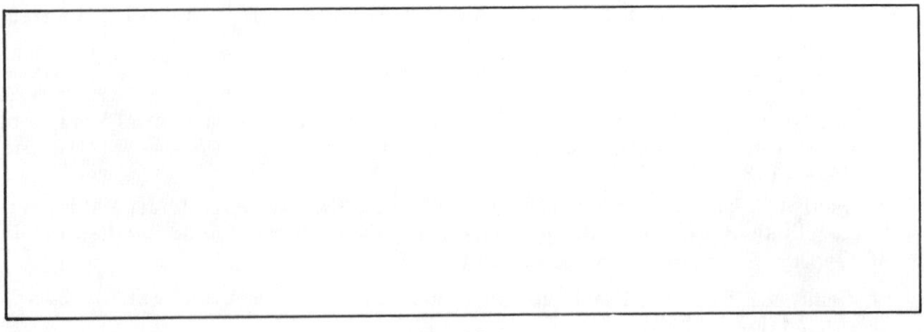

Fig. 10.1

2. Draw a rectangle that is about 1/2 of the one shown.
3. Draw a rectangle that is about 3/4 of the one shown.
4. Draw a rectangle that is about 9/10 of the one shown.

After students have drawn these rectangles, ask the following questions to help them reflect on their work and lead them to some correct understandings:

1. Was each rectangle you drew larger or smaller than the given rectangle?
2. How can you tell whether the shape you draw is going to be a lot smaller or just a little smaller than the given rectangle? Explain.
3. Did the shapes you drew get increasingly larger or smaller? Why?

In my experience it is more productive to summarize and record students' findings before continuing to set B. Leaving this information on the board or overhead projector makes it easier to compare it with later findings.

In set B students find products for exercises in which one factor is a fraction used in set A. Its purpose is to build on the awareness of the comparative sizes of these fractions by finding these fractional parts of the same number.

Set B: Find the products.

1. 1/10 of 3 1/3 = ______
2. 1/2 of 3 1/3 = ______
3. 3/4 of 3 1/3 = ______
4. 9/10 of 3 1/3 = ______

Questions

1. Was each product greater than or less than 3 1/3?
2. How can you tell whether the product is going to be a lot smaller or just a little smaller than 3 1/3?
3. Did the products get increasingly larger or smaller? Why?

Now it is important to compare the results of both sets of exercises. It is also good to reemphasize that we multiply by a fraction (less than 1) to find a *part* of an amount or a number. (To find 1/2 of an amount means that you do *not* want to find *all* of it. Therefore the product must be less than the whole amount.) The discussion also reminds students that multiplication does not *always* produce products that are greater than both factors.

The following activities cause students to reinforce and apply the basic understandings developed in the first two activities. The use of numbers that are awkward to work with makes it desirable to estimate. You can time the activity to prevent students from actually computing the products.

Set C: Estimate to choose the product *closest* to the exact answer.

	a	*b*	*c*	*d*
1. 9/10 of 7 5/6 ≈	7 1/2	3 1/2	8 1/3	15 3/4
2. 1/2 × 22 3/4 ≈	20 1/2	32 3/4	23 1/2	11 3/8
3. 3/20 of 17 1/2 ≈	2 5/8	15 1/2	20 2/3	30 3/4
4. 3/4 × 97 7/8 ≈	96 1/3	73 2/3	20 1/2	121 1/3

Set D: (Challenge problem): Number each item so the products will be in order from smallest to largest.

3/8 × 19 2/3   1/10 × 19 2/3   3/4 × 19 2/3   1/2 x 19 2/3

Up to this point the work in estimating the fractional part of a number has used fractions less than 1. Now students are given similar activities in which the first factor is greater than 1.

Set E: Draw the four rectangles described below:

1. Draw a rectangle that is 1 1/2 times as big as the one in figure 10.2.
2. Draw a rectangle 2 times as big as the one shown.
3. Draw a rectangle 3 1/2 times as big as the one shown.
4. Draw a rectangle 4 times as big as the one shown.

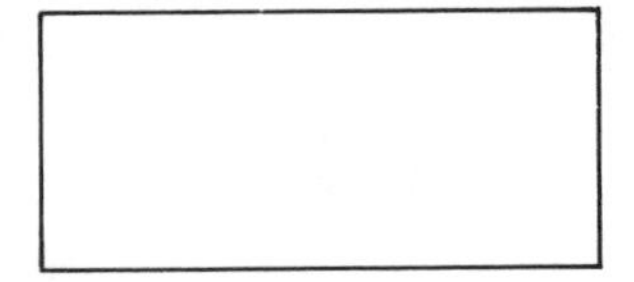

Fig. 10.2

Questions

1. Was each rectangle you drew larger or smaller than the given rectangle?
2. How can you tell whether your rectangle is going to be a lot larger or just a little bit larger than the given rectangle? Explain.
3. Did the rectangles you drew get increasingly larger or smaller? Why?

After these results are discussed and recorded, students can work on set F, which builds on the work of set E.

Set F: Find the products.

1. 1 1/2 × 3 1/3 = _______
2. 2 × 3 1/3 = _______
3. 3 1/2 × 3 1/3 = _______
4. 4 × 3 1/3 = _______

Questions

1. Is each product greater than or less than 3 1/3?
2. How can you tell whether the product is going to be a lot larger or just a little bit larger than 3 1/3?
3. Did the products get increasingly larger or smaller? Why?

Again, it is important to compare the results from set E and set F and also compare these results with those of sets A and B. Students should now be able to estimate when the product will be greater than the second factor, as well as when it will be less than the second factor. Exercise set G gives students the opportunity to apply what they have learned about both types of problems.

Set G: Estimate to choose the product *closest* to the exact answer.

	a	*b*	*c*	*d*
1. 1 1/2 × 4 2/3 ≈	5 1/3	7 1/2	3 3/4	2 1/3
2. 1/2 × 4 2/3 ≈	6 1/3	9 1/2	3 3/4	2 1/3
3. 4 3/4 × 6 1/10 ≈	10 7/8	18 2/3	30 3/4	5 3/40
4. 2 1/10 × 7 7/8 ≈	16	8	4	3 3/4

Set H (Challenge problem): Number each exercise so the products will be in order from smallest to largest.

1 3/8 × 6 1/2   2 2/3 × 6 1/2   1/2 × 5 1/2   9/10 × 5 1/2

SUMMARY

These activities have led to improved reasoning and performance by my students on estimating products of fractions as well as on solving word problems that involve fraction multiplication. The activities are particularly effective when the work is done as a group under my direction rather than independently. I believe this "picture" method of teaching estimation focuses their attention on the relative *amounts* involved and not just the symbols.

My success with this method of teaching the multiplication of fractions has led me to extend the method to division, since students also have difficulty estimating and solving word problems with that operation. Similar exercises with decimals have also proved helpful.

I believe it is very important for students to learn to estimate so they will spot careless errors and be able to answer the critical question, "Is my answer reasonable?" However, I also believe that the thinking and reasoning involved is the same as that used to identify the correct process for solving most word problems. Thus work on estimation has a carry-over benefit for solving word problems. In fact, this additional benefit has probably been the most useful to many of my students.

11

Components of Mental Multiplying

Donald W. Hazekamp

MENTAL arithmetic is the computing of exact answers without paper and pencil or other computational aids, usually with nontraditional mental processes. Mental arithmetic is an important component of computation and estimation, and it aids in problem solving (Spitzer 1967; Trafton 1978). Many situations in daily life require exact answers but are simple enough to be done mentally. Wandt and Brown (1957) noted that nearly half the nonoccupational uses of mathematics were in the area of exact mental calculations.

Put yourself in this situation:

> You have only \$10 to spend. You need 5 notebooks. Each notebook costs \$1.85. You would also like to buy a 35¢ eraser. Do you have enough money?

To solve, you might estimate and think:

$1.85 ≈ $2
5 × $2 = $10
Yes, I have enough for the notebooks.

But, then you don't know if there is enough for the eraser. Estimating isn't good enough. You need to find the exact cost of the notebooks. You think further and calculate:

$2 − $1.85 = 15¢
5 × 15¢ = 5 × 10¢ + 5 × 5¢
= 50¢ + 25¢ = 75¢
$10.00 − 75¢ = $9.25, the exact cost of 5 notebooks

Or, alternatively, perhaps this is the way your thinking goes:

5 × $1.85 5 × $1 = $5.00
5 × .80 = $4.00
5 × .05 = .25
9.25 exact cost of 5 notebooks

Then to complete the problem you think:

$9.25 .35 $9.60	$10 − $9.60 = 40¢, which is enough for the 4% sales tax Yes, the $10 is enough.

Although the problem is easy enough to be done mentally, many students and adults would not be able to solve it without paper and pencil or a calculator (on which they would have to calculate two or three times before being sure of their answer). This is because they lack instruction in mental computation. Flournoy (1957, 1959) and Atweh (1982) have both noted the lack of this type of instruction. Teachers need to include work with oral or unwritten computation in their arithmetic instruction.

Many mental computation situations involve multiplication. Several methods for mental multiplying have been devised. Many are very elaborate, particularly those for multiplying large numbers. Not much is known about the way students develop these thinking skills, and probably no one technique for performing mental multiplication can be called the best. However, it is evident that students do gain when systematic instruction is given. It appears that gains are made when various approaches are used and students adopt the ones that seem best for them.

This article focuses on some of the techniques used in finding exact products mentally. The discussion is limited to simple multiplication situations that occur in daily life and will begin with special products, the essential component of mental multiplication. Then two strategies for mental multiplying are described: *front-end* and *compensation.* Requisite concepts and skills are noted. And, finally, some suggestions are given for teaching special products and mental multiplying.

SPECIAL PRODUCTS: AN ESSENTIAL COMPONENT

In all mental multiplication, understanding the situation and having the skill of finding *special products* are as important as knowing the basic multiplication facts. The term broadly refers to those products that are easily found by multiplying by a power of 10 or a multiple of a power of 10. Some of them are illustrated below.

1. Ones times tens, hundreds, thousands, and so forth:

   $8 \times 30 = 240$, or $30 \times 8 = 240$
   $6 \times 500 = 3\ 000$, or $500 \times 6 = 3\ 000$
   $7 \times 4\ 000 = 28\ 000$, or $4\ 000 \times 7 = 28\ 000$

2. Tens, hundreds, and thousands multiplied together:

   $40 \times 80 = 3\ 200$
   $50 \times 600 = 30\ 000$
   $800 \times 2000 = 1\ 600\ 000$

3. Tens, hundreds, or thousands times two- or three-digit numbers:
   $30 \times 24 = 720$ $4\ 000 \times 62 = 248\ 000$

THE FRONT-END APPROACH

The front-end approach to multiplying mentally is characterized by renaming (or estimating) one of the factors, multiplying the parts separately, and then adding (or subtracting). The following examples illustrate:

1. 5×38

THINK

$38 = 30 + 8$
$5 \times 30 = 150$
$5 \times 8 = 40$
$150 + 40 = 190$
$5 \times 38 = 190$

2. 5×38

Round 38 to 40
$5 \times 40 = 200$
$40 - 38 = 2$
$5 \times 2 = 10$
$200 - 10 = 190$
$5 \times 38 = 190$

These techniques can also be used for special products, as shown in example 3:

3. 400×73

$73 = 70 + 3$
$400 \times 70 = 28\ 000$
$400 \times 3 = 1\ 200$
$28\ 000 + 1\ 200 = 29\ 200$
$400 \times 73 = 29\ 200$

Note that the front-end approach is just as systematic as the traditional algorithm but makes it easier to remember the partial products. It has been shown that this method is often used by good estimators (Reys and Bestgen 1981). Another method (somewhat less useful) is the compensation approach.

THE COMPENSATION APPROACH

In the compensation approach, the product is found by using easy multiplications and divisions to simplify the problem. This is particularly useful when one of the numbers is a factor of a power of 10. The following examples illustrate:

1. Multiply one factor and divide the product by the same number.

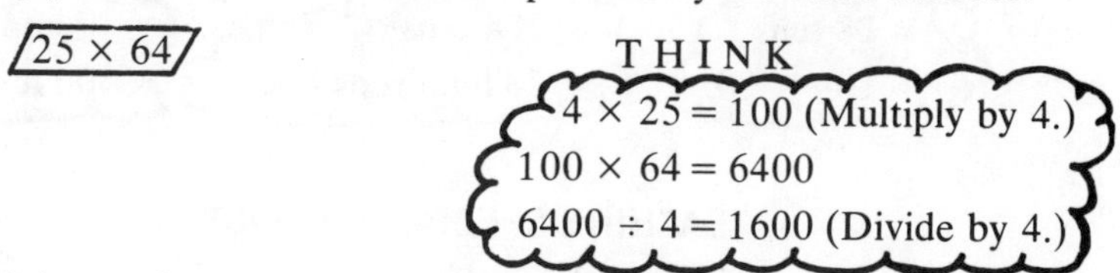

2. Multiply one factor and divide the other factor by the same number.

3. Successively double and halve each factor until one of them is a power of 10.

4. Rules to multiply by 5, 50, 500, and so forth:

   By 5: Divide by 2, multiply by 10. — 5 is 10 ÷ 2

   By 50: Divide by 2, multiply by 100. — 50 is 100 ÷ 2

   By 500: Divide by 2, multiply by 1000. — 500 is 1000 ÷ 2

   364 × 50

   364 ÷ 2 = 182
   182 × 100 = 18 200

NEEDED CONCEPTS AND SKILLS

Certainly the knowledge of basic multiplication facts is a prerequisite for good mental multiplying. But an examination of the preceding examples shows that mental multiplication, more importantly, probably requires a better understanding of concepts and more use of special multiplying skills than the written paper-and-pencil algorithm does. For example, the front-end approach involves an implicit use of both the left and right distributive properties; the compensation approach involves an understanding of inverse operations. Both approaches, as well as special products, involve the need for a good understanding of decimal numeration concepts. Such an understanding will give students (1) the ability to recognize and work with numbers that are multiples of powers of 10, and (2) a mental flexibility that will allow them to think of numbers in many different ways and forms: ones, groups of tens, groups of hundreds, and so forth.

For example:

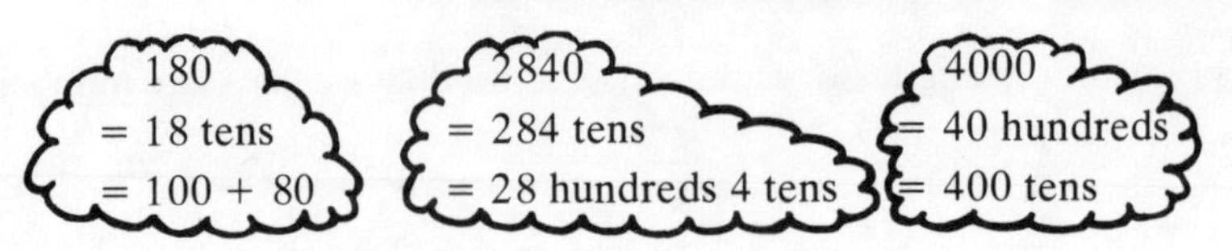

TEACHING SPECIAL PRODUCTS

Special products need to be taught carefully. Because round numbers are easy to work with, it is easy for pupils to adopt rote procedures that can produce incorrect answers and little understanding of the number properties involved. For example, when multiplying 50 × 60, pupils often get 300 rather than 3000; they reason that the product can have only two zeros. Understanding the concepts involved and gaining skill in finding special products are necessary before students can effectively apply them in mental multiplying situations. Instruction using base representations for numbers has been found effective in teaching special products (Hazekamp 1977).

In all work with products, a knowledge of basic multiplication facts is essential. But this does not mean that the teaching of special products must be delayed until all facts through the nines are learned. Activities that establish the patterns (ones × tens = tens; ones × hundreds = hundreds; ones × thousands = thousands; tens × tens = hundreds, etc.) can be started when students have mastered only the simpler facts, such as 2 × 3 = 6, 3 × 4 = 12, or 4 × 2 = 8. The key is to teach the special products that require only those basic facts the pupils have learned. Some suggestions for teaching special products follow.

1. Use an array model to illustrate the product of ones × tens:

   Rename 12 tens as 120: so *3 × 40 = 120.*

   Then, present and assign written exercises:

   3 × 6 = *18*, 3 × 6 tens = *18* tens
   so 3 x 60 = 180,

   $$\begin{array}{r}7\\ \times 4\\ \hline 28\end{array} \quad \begin{array}{r}7\text{ tens}\\ \times 4\quad\ \\ \hline 28\text{ tens}\end{array} \quad \text{so} \quad \begin{array}{r}70\\ \times 4\\ \hline 280\end{array}, \quad \text{and finally} \quad \begin{array}{r}80\\ \times 4\\ \hline 320\end{array}$$

   Then use oral exercises to develop the following thinking pattern:

   6 × 40

   40 = 4 tens
   To show tens I need 0 in ones place.
   Multiply 6 × 4 to find the number of tens.
   24 tens is *240*.

   Finally, give similar exercises for multiples of a hundred and a thousand:

a)	2 hundreds	200	*b)*	6 thousands	6 000
	× 6	× 6		× 3	× 3
	12 hundreds	1200		18 thousands	18 000

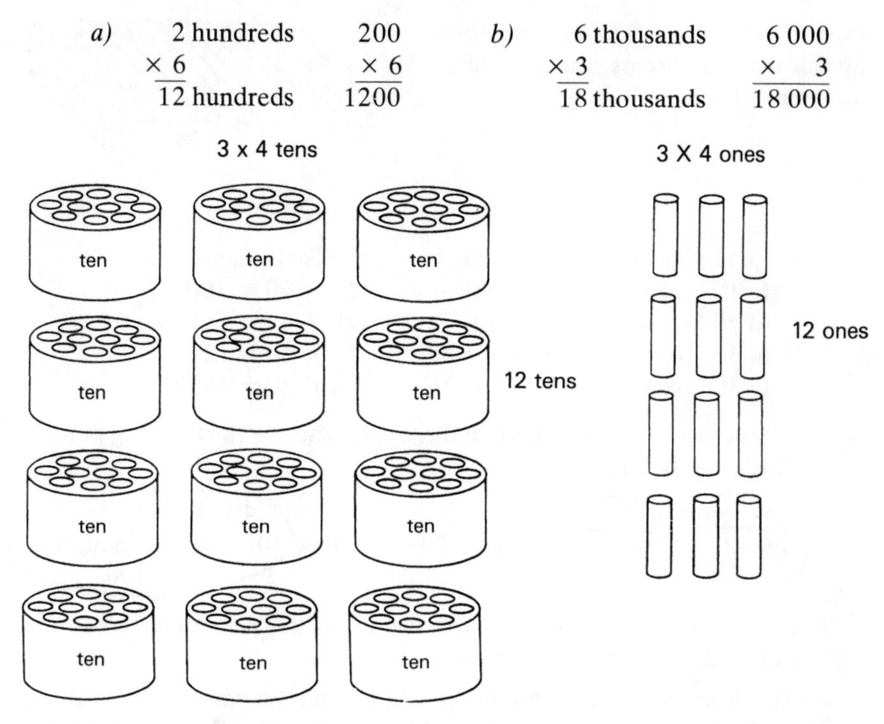

2. Students often have place-value difficulties with exercises like 5 × 800 where they get a product of 400 instead of 4000. Thus exercises such as the ones below need to be included and given particular attention.

a)	5 tens		50	*b)*	4 thousands		4 000
	× 6	so	× 6		× 5	so	× 5
	30 tens		300		20 thousands		20 000

3. For multiplying tens by tens, illustrations that extend the concept of 10 tens = 1 hundred are effective to establish the pattern m tens × n tens = $(m \times n)$ hundreds.

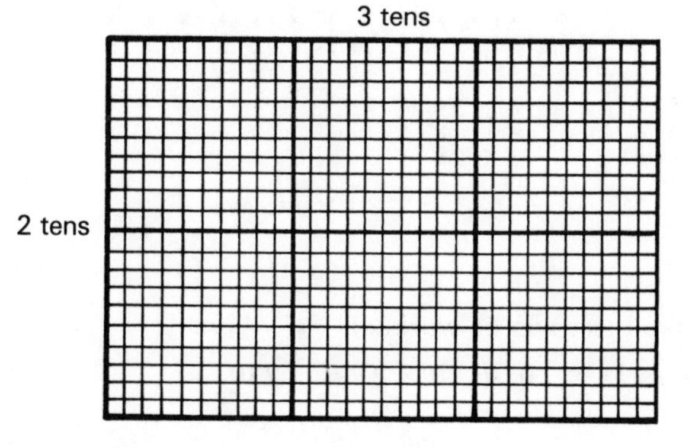

(2 tens) x (3 tens) = 6 hundreds

so **20** x **30** = **600**

Next, to develop a firm understanding of this pattern, discuss situations like the one below:

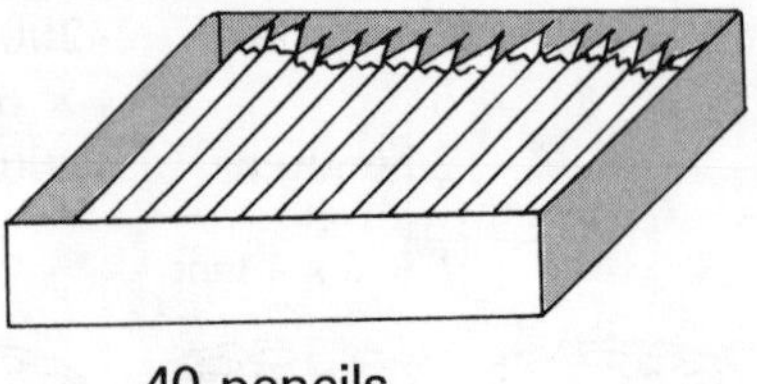

40 pencils

Each box has 40 pencils. How many pencils are there—

In 10 boxes?	(10 sets of 40)	**10** × **40** = **4**00
In 20 boxes?	(20 sets of 40)	**20** × **40** = **8**00
In 30 boxes?	(30 sets of 40)	**30** × **40** = **12**00
In 40 boxes?	(40 sets of 40)	**40** × **40** = **16**00

Discussion of examples like these can generate the thinking pattern illustrated below:

(4 tens) (8 tens)
40 × 80 = ? 40 × 80 = ? hundreds — (4 × 8 = *32*
40 × 80 = *32* hundreds
= 3200)

Later you can demonstrate extensions to the pattern m tens × n hundreds = ($m \times n$) thousands.

4. The order in which the multiplication is done can make a difference to students. For example, for 10 × 13 they often give the response 1013 instead of 130. This can even happen after a correct response of 13 × 10 = 130 has been given. An array (fig. 11.1) is helpful for correcting this. Turning the array (fig. 11.2) shows that 10 thirteens is the same as 13 tens (10 × 13 = 130).

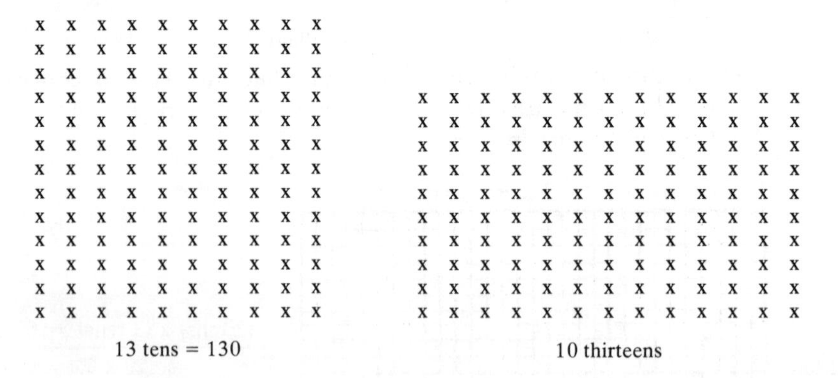

Fig. 11.1 Fig. 11.2

Then more exercises like these can be worked:

- 10 × 23 is the same as 23 × 10, so 10 × 23 = 230

- 27 × 100 = 2700, so 100 × 27 = 2700
- 10 × 16 = 160; 100 × 62 = 6200; 38 × 100 = 3800
- 1000 × 43 = 43000.

As the previous examples show, the teaching of special products needs to highlight thinking of numbers as groups of tens, hundreds, thousands, and so forth. Furthermore, to increase their proficiency, teach students how to find special products by a simple one-line procedure before teaching them the exact computational algorithms for two- and three-digit numbers. Doing this helps prevent them from applying paper-and-pencil algorithmic procedures to simple special-product exercises.

In multiplying 600 × 41, for example, students should be able to find the special product by writing—

$$600 \times 41 = 24\ 600 \quad \text{or} \quad \begin{array}{r} 600 \\ \times\ 41 \\ \hline 24\ 600 \end{array}$$

Develop this by trying to establish the following type of thinking pattern:

Students should *not* use such algorithmic forms as these:

$$\begin{array}{r} 41 \\ \times\ 600 \\ \hline 00 \\ 00 \\ 246 \\ \hline 24{,}600 \end{array} \quad \text{or} \quad \begin{array}{r} 600 \\ \times\ 41 \\ \hline 600 \\ 2400 \\ \hline 24{,}600 \end{array}$$

These procedures not only lead to the likelihood of place-value errors but cannot be readily done mentally.

SOME SUGGESTIONS FOR TEACHING MENTAL MULTIPLICATION

Mental arithmetic suggests, first of all, the need for oral work. Oral arithmetic has benefits at all grade levels because it eliminates reading and writing difficulties and also focuses students' attention on the number relationships involved. To develop proficiency in this area, an ongoing program is needed in which a minimum of five to fifteen minutes two or three times a week is spent on oral activities. In the beginning, students who have not had this type of work will show a natural reluctance, but the program should gradually remove the need for paper and pencil, particularly for students who have worked only with the written algorithm.

Second, since mental arithmetic aids consumers in daily computations and problem solving, it should be taught in the context of real-world applications.

Third is the major question of its placement in the instructional program. Currently, if mental arithmetic is taught at all, it follows the exact written computational procedures. As a result, many pupils, particularly those with good written computational skills, attempt to solve the problems mentally using exact written computational techniques. For example, some will think like this:

But they get lost or make errors in regrouping when they do not have pencil and paper; usually the exact written format is too difficult to do mentally. Special mental multiplying procedures need to be taught early and integrated into the multiplication program before students have almost mastered the written algorithm.

A good time to begin is after they have had some experience with the distributive property and have worked with special products. The teacher's goal is to show students how to find the product of, for example, 6×43 by thinking:

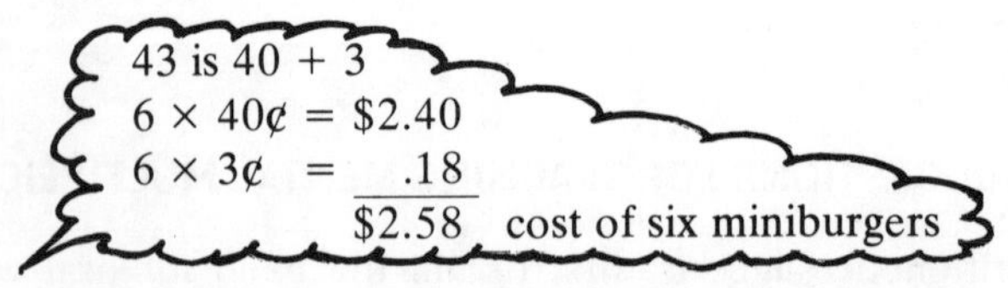

How can this goal be reached? For the first few examples, discuss and write on the chalkboard the thinking process used. Then practice with other examples, such as 7×35, thus:

Teacher's question	Student's response	Write only the partial products
"What do we multiply first?"	"7 × 30"	210
"What do we multiply next?"	"7 × 5"	35
"Now what is 7 × 35?"	"245"	

For the first few exercises, let students write the partial products on paper. Then on the last few, have them write only the final answer.

Writing down the partial products would go something like this.

Exercise	**Students write**
4 × 76	280 + 24 = 304
50 × 46	2000 + 300 = 2300

Ask the class if anyone got the answer a different way. For 4 × 76 some might say

$$76 = 75 + 1 \qquad 2 \times 75 = 150$$
$$4 \times 75 = 300,$$
and 4 more is 304.

For 50 × 46 others might say

$$50 \times 2 = 100 \qquad 46 \div 2 = 23$$
$$23 \times 100 = 2300.$$

Finally be sure to include a variety of everyday examples:

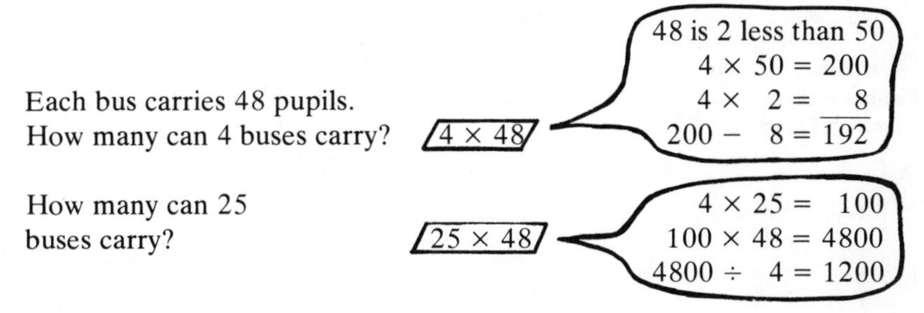
48 is 2 less than 50
4 × 50 = 200
4 × 2 = 8
200 − 8 = 192
Each bus carries 48 pupils.
How many can 4 buses carry?
4 × 48
How many can 25
buses carry?
25 × 48
4 × 25 = 100
100 × 48 = 4800
4800 ÷ 4 = 1200

Each box contains two dozen pencils. Will 25 boxes be enough to give 640 pupils one pencil each?

25 × 24 is the same as
50 × 12 is the same as
100 × 6 = 600 (not enough)
2 × 24 = 48, 600 + 48 = 648
We will need 2 more boxes.

SUMMARY

In this presentation of some of the basic components of mental multiplying, the focus has been on (*a*) the role of special products and (*b*) procedures for developing understanding as well as computational proficiency.

Special products, besides their use in exact mental multiplication, are useful in product estimation and mental division. Two techniques for finding exact products have been described, front-end and compensation. It is hoped that they, along with the accompanying teaching suggestions and exercises, will give teachers and students a good start in mental computation. Such skills are developed only with regular practice throughout the school year. You are encouraged to use these ideas as only a beginning and to venture out trying other approaches suggested by your students or gleaned from outside the regular textbook.

REFERENCES

Atweh, Bill. "Developing Mental Arithmetic." In *Mathematics for the Middle Grades (5–9)*, 1982 Yearbook of the National Council of Teachers of Mathematics, pp. 50–58. Reston, Va.: The Council, 1982.

Flournoy, Frances. "Developing Ability in Mental Arithmetic." *Arithmetic Teacher* 4 (October 1957): 147–50.

———. "Providing Mental Arithmetic Experiences." *Arithmetic Teacher* 6 (April 1959): 133–39.

Hazekamp, Donald W. "The Effects of Two Initial Instructional Sequences on the Learning of the Conventional Two-Digit Multiplication Algorithm in Fourth Grade." (Doctoral dissertation, Indiana University, 1976.) *Dissertation Abstracts International* 37 (19u7): 4933.

Reys, Robert E., and Barbara J. Bestgen. "Teaching and Assessing Computational Estimation Skills." *Elementary School Journal* 82 (November 1981): 117–27.

Spitzer, Herbert F. *Teaching Elementary School Mathematics.* Boston: Houghton Mifflin Co., 1967.

Trafton, Paul R. "Estimation and Mental Arithmetic: Important Components of Computation." In *Developing Computational Skills,* 1978 Yearbook of the National Council of Teachers of Mathematics, pp. 196–213. Reston, Va.: The Council, 1978.

Wandt, Edwin, and Gerald W. Brown. "Non-Occupational Uses of Mathematics." *Arithmetic Teacher* 4 (October 1957): 151–54.

12

Estimating Decimal Products: An Instructional Sequence

James H. Vance

THIS article outlines an instructional sequence for developing the understanding and skills needed to estimate, and check for the correctness of, answers when multiplying two decimal numbers.

The unit of instruction grew out of a study assessing the ability of seventh- and eighth-grade students to compute and estimate the products of two decimal numbers (Vance 1983). At each grade level the test papers of ten boys and ten girls at each of three achievement levels were analyzed. In addition, individual interviews were conducted with eight students from each grade who were in the same school but did not take the paper-and-pencil test. Following the testing and interviews, the investigator taught a series of lessons on estimating decimal products to a seventh-grade class in another school. The unit was subsequently revised and extended.

SEQUENCE OF SKILLS

The skills are organized under six headings and are stated as performance objectives. An example illustrates each objective.

1. *Place Value*

   a) Read and write a multidigit number.

   The numeral 700 020 006.018 is read, "seven hundred million, twenty thousand, six, and eighteen thousandths."

   b) Identify place value in a multidigit number.

   In 694 702.135, the 9 is in the *ten thousands* place and means *90 000* ("ninety thousand"). The 3 is in the *hundredths* place and means 0.03 ("three hundredths").

2. *Order and Comparison*

   a) Continue counting sequences.

   4.07 4.08 4.09 4.10 4.11

b) Identify equivalent decimal forms.

0.6 (underlined) 0.06 0.60 (underlined) 0.006 0.600 (underlined)

c) Find a number between two given numbers.

0.72 0.721 (underlined) 0.73

d) Compare two numbers.

0.7 $\underline{\ >\ }$ 0.68

e) Order a set of numbers from least to greatest.

3.461 3.61 3.64

3. *Leading-Digit and Rounded Leading-Digit Estimates*

a) Write the leading-digit (or front-end) estimate for a number.

7964.913 → 7000
0.0982 → 0.09

b) Write the rounded leading-digit estimate for a number.

7964.913 → 8000
0.0982 → 0.1

4. *Multiplication by Powers of 10*

a) Multiply a number by 10, 100, 1000.

$1000 \times 3.2 = 3200$

b) Multiply a number by 0.1, 0.01, 0.001.

$8.46 \times 0.01 = 0.0846$

c) Find the product of two powers of 10.

$1000 \times 0.01 = 10$

d) Find the product of two multiples of powers of 10.

$40 \times 0.008 = 0.32$

5. *Estimating Products*

a) Use leading-digit estimators.

$862 \times 0.36 \rightarrow 800 \times 0.3 = 240$

b) Use rounded leading-digit estimators.

$862 \times 0.36 \rightarrow 900 \times 0.4 = 360$

c) Use the "round up, round down" strategy.

$862 \times 0.36 \rightarrow 900 \times 0.3 = 270$

6. *Fraction Estimators*

a) Recognize the decimal equivalent of a simple fraction.

$1/4 = 0.25$

b) Identify decimals that have simple fraction estimators.

0.261 → 1/4

c) Find the product of a fraction and a whole number or decimal.

$1/4 \times 2800 = 700$

d) Estimate products using a fraction estimator and a "compatible" estimator.

$0.261 \times 2734.5 \rightarrow 1/4 \times 2800 = 700$

7. *Recognizing Incorrect Anwers*

a) Identify unreasonable answers.

$34 \times 62 = 218 \qquad (30 \times 60 = 1800)$

b) Identify answers that are too high or too low.

$34 \times 62 = 1798$

$34 \times 62 = 1808 \qquad (30 \times 60 + 4 \times 2 = 1808)$

c) Identify answers with a wrong final digit.

$34 \times 62 = 2106 \qquad (4 \times 2 = 8)$

d) Identify answers with a misplaced decimal point.

$358.6 \times 0.25 = 8.965 \quad (400 \times 0.2 = 80; \text{ or } 360 \times 1/4 = 90)$

BACKGROUND AND SUGGESTED ACTIVITIES

Each of the seven skills listed above will now be discussed further.

Place Value

The ability to estimate products rests ultimately on a firm understanding of place-value ideas. Concrete models must be provided and oral language using place-value names must be emphasized to give students some feel for the quantities represented by large and small numbers. As evidence of the need for oral language, consider students who, when interviewed in the study and asked to read the numeral 2.3, always said, "two point three" or "two decimal three." Few, even when pressed, said "two and three-tenths." Base-ten blocks, which are used in the early grades to represent ones, tens, hundreds, and thousands, can also be used to model ones, tenths, hundredths, and thousandths (fig. 12.1).

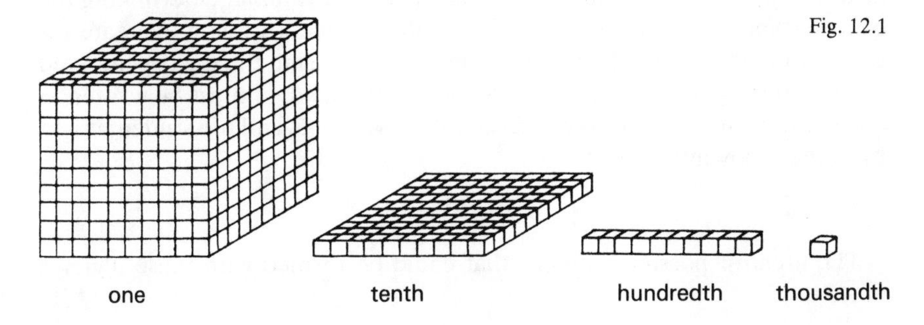

Fig. 12.1

There are two important ideas in decimal numeration: (1) the value of any position is ten times the value of the position to its immediate right; and (2) the decimal point determines the ones place. The central role of the ones position and the relationships among the various positions are indicated in the place-value chart in figure 12.2.

thousands	hundreds	tens	ones	tenths	hundredths	thousandths
6	9	4	3	7	2	5

6943.725

Fig. 12.2

Another device that can help learners appreciate the value of each digit in a number is the hand-held calculator. For example, a student can enter 1324.759 and then attempt to wipe out the 3 or 5 in a single operation. (This is accomplished, of course, by subtracting 300 or 0.05.) Another task is to enter a number such as 2060.807 from right to left (0.007 + 0.8 + 60 + 2000).

Order and Comparison

Ordering and comparing activities provide further opportunities for the students to apply and extend place-value concepts. One important idea underlying these skills is that a decimal fraction has many names: for example, 0.3 = 0.30 = 0.300. Although most students interviewed could identify the equivalent forms of 0.3, few could explain why zeros can be annexed in this way. Only one individual noted that 3/10 = 30/100. Students also had difficulty naming a number between 0.72 and 0.73. Some stated that there wasn't one, whereas others suggested 0.72 1/2. Ordering such numbers as 0.7 and 0.68 involves this same concept. Both the blocks and the place-value chart can be used to model the relationship (fig. 12.3).

The following game gives practice in ordering numbers and reinforces the idea that place value is more important than face value in determining the contribution of a digit to the size of a number. Players attempt to create the greatest number with a certain number of places using the digits generated through tossing a die. With each roll of the die the digit must be written in one of the positions. For example, on rolls of 4, 1, 3, 5, and 6, a student might form the following number:

$$\underline{5}\ \underline{4}\ .\ \underline{3}\ \underline{6}\ \underline{1}$$

The greatest possible number that could be formed with these digits is 65.431.

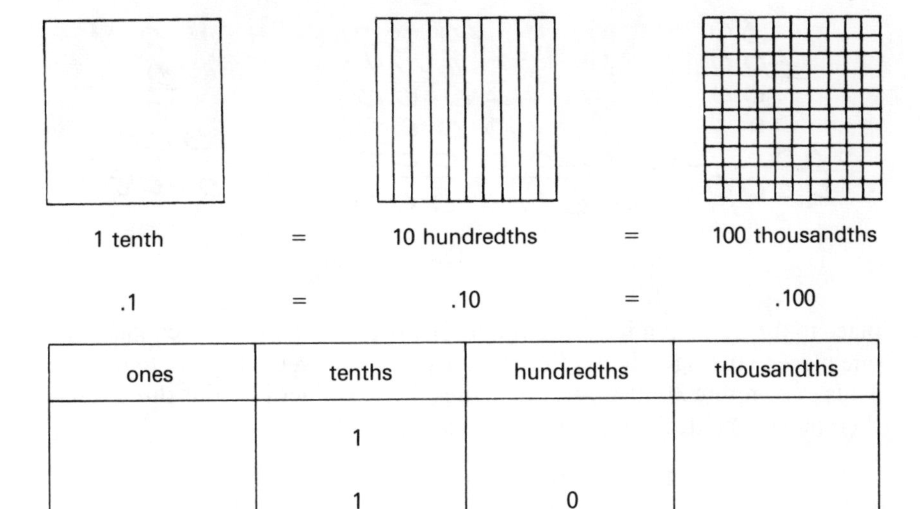

ones	tenths	hundredths	thousandths
	1		
	1	0	
	1	0	0

Fig. 12.3

Leading-Digit and Rounded Leading-Digit Estimators

Students need to appreciate that the leading (nonzero) digit in a multidigit number is the important one in determining the "size" of the number. Replacing the other digits by zeros (or "dropping" them) and then reading the resulting number orally is a valuable exercise. This skill should be mastered by students before they attempt to round the leading digit. It should be pointed out that the leading-digit estimate is always less than or equal to the actual number whereas the rounded leading-digit estimate can be either greater or less than the number.

Multiplication by Powers of 10

The computation test included seven items in which both factors were numbers with a single nonzero digit (e.g., 0.06×0.009) or one factor was a power of 10 (e.g., 10×6.14). An examination of pupil work revealed that about 40 percent of these exercises were attempted using a long algorithm with unnecessary zeros. Three examples of student work are shown in figure 12.4. (Exercises were presented in horizontal form.)

Rules for multiplying by a power of 10 follow from a consideration of the place-value chart. For example, multiplying a number by 10 "moves" each digit one place to the left. This is the same as moving the decimal point one

50	10.00	0.4
600	6.14	0.2
00	40000	08
00	10000	000
300	600000	0.08
30000	61.4000	

Fig. 12.4

place to the right. Similarly, multiplying by 0.1 (one-tenth) moves each digit one place to the right; hence the decimal point is moved one place to the left. Thus to compute 3000 × .07, one writes the basic fact 21, adds three zeros, then moves the decimal point two places to the left:

$$21 \rightarrow 21000 \rightarrow 210.00 \rightarrow 210$$

Efficient procedures for multiplying by powers and multiples of 10 were emphasized during the classroom piloting of the unit. However, a test given a week after the instruction disclosed that those students who tended to use an inefficient algorithm at the beginning of the unit continued to do so. When questioned, these students said that they felt more secure using the standard algorithm for all multiplication exercises, even though they were aware that in some examples certain zeros or rows of zeros could be omitted and in others a direct method not involving a vertical form could be used.

Estimating Products

The multiple-choice estimation test included seven items that were parallel to the computation exercises described in the previous section. Performance on the computation questions was much better than on the corresponding estimation questions in which students were only required to select the answer with the correct order of magnitude. For example, 82 percent of the subjects computed the correct answer to 0.4 × 0.2, but on the related estimation items, 0.378 × 0.24, only 41 percent chose the alternative with the decimal point correctly positioned relative to the digit 8. (About 20 percent of the students chose each of the other alternatives: 0.00008, 0.008, 0.8.) It was found that students who obtained the correct answer to a computation question using an efficient procedure scored about 9 percent higher on the corresponding estimation item than students who obtained the correct answer using an inefficient algorithm.

Students need to be taught to use the leading-digit or rounded leading-digit estimator to estimate products; they do not do this automatically. For example, when asked to estimate 48.52 × 623.8, many students rounded to the nearest whole number (49 × 624) or rounded both numbers to the nearest ten (50 × 620). Reluctance or inability to round to the leading digit was also evident for 6.825 × 0.012. Common responses were 7 × 0 or 6.8 × 0 (producing an estimate of zero) and 6.83 × 0.01. Many students believed that both factors should be rounded to the same number of decimal places.

After students have had considerable practice in estimating products using leading-digit and rounded leading-digit estimators, the "round up, round down" strategy can be introduced. For example, a better estimate for 25 × 35 than either 600 (20 × 30) or 1200 (30 × 40) would be 800 (20 × 40) or 900 (30 × 30). "Number sense" in estimation (Johnson 1979) develops with experience and maturity.

Fraction Estimators

In certain situations, the process of computing or estimating the product of two decimals can be simplified by expressing or approximating one of the factors by a simple fraction. For example, 360 × 0.25 can be thought of as one-fourth of 360. In the study, only 3 of 120 students computed the answer to this question in this way.

To begin using fraction estimators, students need to be able to recognize the decimal equivalents of 1/2, 1/3, and 1/4 and identify decimals that are "close to" these fractions. Once a fraction estimator has been determined, the estimate for the other factor is chosen so that it is divisible by the denominator of the fraction. This "compatible" number may have either one or two nonzero digits. For example, 276 × 0.35 can be estimated as 1/3 of 270.

Recognizing Incorrect Answers

The first six sections of the sequence describe the skills needed to find an approximate answer for a multiplication question and perhaps determine whether the estimate is higher or lower than the actual product. In this final section, ways of recognizing answers that are close but not exact are considered. An understanding of the multiplication algorithm and of the significance of each partial product is required for these skills, as illustrated in figure 12.5.

```
  62
  34
----
   8    (4 × 2)   = last digit in the answer
 240    (4 × 60)
  60    (30 × 2)
1800    (30 × 60)= leading-digit estimate (too low)
----
```

Fig. 12.5

In refining their estimates, many students interviewed gave answers that indicated they were unaware of the middle terms in the algorithm. For example, an "estimate" given for 6.138×10.35 was 60.56 ($6 \times 10 + .14 \times .4$), and no further refinement was made. Many students were also unable to predict the final digit in the product of two numbers and therefore did not recognize situations in which a trailing zero after the decimal point had been omitted. In one estimation item, students were asked to place the decimal point in the answer to the following exercise:

$$358.6 \times 0.25 = \boxed{8 \quad 9 \quad 6 \quad 5}$$

Eighty percent chose 8.965 as the answer. They seem to have counted over three decimal places without realizing that the final digit, zero, has been dropped.

This problem yielded another interesting result. Toward the end of each individual interview, the student was asked to estimate 358.6×0.25. After a satisfactory estimate was obtained, the subject was shown the question in the form just described and asked to place the decimal point in the answer. Invariably the student would count the decimal places in the two factors and give the answer 8.965. The student would then be shown the estimate he or she had previously obtained for the same factors and again was asked to place the decimal point. Most students acknowledged that 89.65 would be closer to their estimate. However, when pressed to decide which of the answers was correct, they would usually place more confidence in the rule for placing the decimal point than in their estimate. The students would then be asked to do the problem on a calculator and consider the answer. Two eighth-grade students suggested that a trailing zero had been dropped, but others remained puzzled when asked for an explanation.

SUMMARY

The process of learning to estimate products, recognize incorrect answers, and reason quantitatively in decimal multiplication situations is a gradual one. The ideas are complex and should be developed in the curriculum over many years. In the middle grades, the teacher can bring the components together, showing how they are related and applied. An instructional unit on estimating decimal products can give students a deeper understanding of fundamental ideas and a greater appreciation of the power and use of mathematics.

REFERENCES

Johnson, David C. "Teaching Estimation and Reasonableness of Results." *Arithmetic Teacher* 27 (September 1979): 34–35.

Vance, James H. *Diagnosing Pupil Performance in Decimal Multiplication: Computation and Estimation.* Report No. 83:13. Vancouver, B.C.: Educational Institute of British Columbia, 1983.

13

Using Money to Develop Estimation Skills with Decimals

Ann C. Kindig

WHEN I began teaching estimation to my middle school students about four years ago, I never anticipated the good results it would produce. The more able students learned to estimate the answers to computations efficiently, but that was expected. What I didn't expect was that slow learners, too, would become proficient estimators—and that while they were developing estimation skills, their paper-and-pencil computation would improve as well. I also found that students were approaching problem solving more thoughtfully. This became evident when I overheard two algebra students talking about a problem. One was observing, "Well, I know you aren't right because your answer isn't reasonable."

I teach estimation throughout the year and with a variety of topics. However, this article describes the methods I use for teaching estimation with decimal numbers. My sequence for teaching computation and estimation with decimals is somewhat unusual. Before discussing techniques of operating with decimals, I teach students how to estimate mentally. I found that as a result of teaching estimation first, students made far fewer errors in their paper-and-pencil computation than when I taught computational procedures first. A second feature of this program for teaching decimals is a reliance on students' familiarity with money.

There are some prerequisite skills, though, that students must master before they begin to work with estimating in decimal computation exercises. They must, first of all, master estimation skills with whole numbers. Early in my career I assumed that the more able students, at least, would know how to estimate with whole numbers. But I was wrong. So now all my students begin by studying whole-number estimation. I teach rounding and a variety of mental calculation procedures for each of the four fundamental operations. Then on to decimals. The prerequisite skills for decimal estimation are understanding place-value concepts and having a good feeling for the relative size of numbers. I use number lines extensively to develop this latter intuition. In particular, I have students do a lot of activities that require them to determine between which two whole numbers a decimal number lies, and to which of these it is closer.

ADDITION AND SUBTRACTION WITH DECIMALS

When you introduce the estimation of sums and differences involving decimals, let all examples involve numbers greater than 1. Teach students to round to the nearest whole number and then use the procedures they have learned for estimating sums and differences of whole numbers. For large numbers, a second rounding—to multiples of 10 or 100—may be needed. During these lessons, I watch carefully to make sure no one uses a pencil to do their calculations.

After students feel secure with these estimation techniques, you can introduce procedures for making more refined estimates. To do this, I use the students' ability to think in terms of money. Numbers are rounded to the nearest hundredth and "changed" to money. For example, 5.3792 + 6.859 becomes \$5.38 + \$6.86 or, for easier calculation, \$5.40 + \$6.90.

After students have learned how to transform an addition exercise into a "money problem," they are also taught the procedure of "making change" to get better subtraction estimates. When they are working an exercise such as 4.9867 − 2.561, I have them think of the problem as \$4.99 − \$2.56, or \$5.00 − \$2.56. Then they proceed to "make change" by thinking: "It takes 4 cents to make \$2.60; and another 40 cents to make \$3.00; and 2 dollars to make \$5.00; so the change would be \$2.44." The difference, 4.9867 − 2.561, is about 2.44. The student might come even closer by recognizing that the answer to \$4.99 − \$2.56 is one cent less, or \$2.43.

I encourage students to use this process of making change outside of school, too. I suggest that when they are shopping, they should try to figure out how much change they are going to receive before the cash register tells them. Students do, in fact, try this and report the results. One student said he was so surprised when he beat the register that he forgot to pick up his purchase. Parents like this method of decimal instruction because they can see a relationship between what their child does in school and the world outside.

MULTIPLICATION WITH DECIMALS

The money analogy also helps when teaching estimation involving the multiplication of decimals. For purposes of this estimation strategy, the repeated-additions interpretation of multiplication is useful. I explain that for the product 53 × 75, for example, they should think "fifty-three of the 75s."

Initially I have students do some very rough estimating in which they simply decide if the product is going to be larger or smaller than the second factor. For example, for the product 4.2 × 1.98, we talk about the fact that the answer will be greater than 1.98 because there are more than four 1.98s. However, if the exercise is 0.42 × 1.98, students recognize that the answer is less than 1.98 because 0.42 is less than 1. After students become fairly confident with these rough estimates, a more refined estimation procedure is introduced. For products having at least one factor close to 1 or greater than 1, the procedure is as follows:

Procedure	**Example**
1. Order the factors so that the first factor is greater than 1 or if less than 1, rounds to 1.	49.8 × 0.783
2. If the first factor has two or more nonzero digits, round to the nearest whole number, multiple of 10, or multiple of 100, whichever applies.	50 × 0.783
3. Round the second factor to the nearest hundredth; then interpret it as money.	50 × $0.78
4. Reinforce the repeated-addition concept of multiplication while eliciting an estimate from the class.	Ask, "If you had 50 piles of 78¢ each, about how much would you have?" Acceptable answers would be "A little less than $50" ($0.78 is less than $1) or "A little less than $40" (round $0.78 to $0.80). Mental calculators may say "$39."

Students have great success with this method. I avoid examples in which both factors are very small, such as 0.054 × 0.031. My students learn to estimate with factors like these when they study scientific notation.

DIVISION OF DECIMALS

The first lessons on the division of decimals are similar to the introduction to multiplication in the sense that they, too, deal with very rough estimates. Students are asked to recognize only if a quotient is greater or less than 1. For the problem 0.5123 ÷ 1.2, a student might pose the question, "How many 1.2s can I take out of 0.5?" Then they can see that the answer must be less than 1.

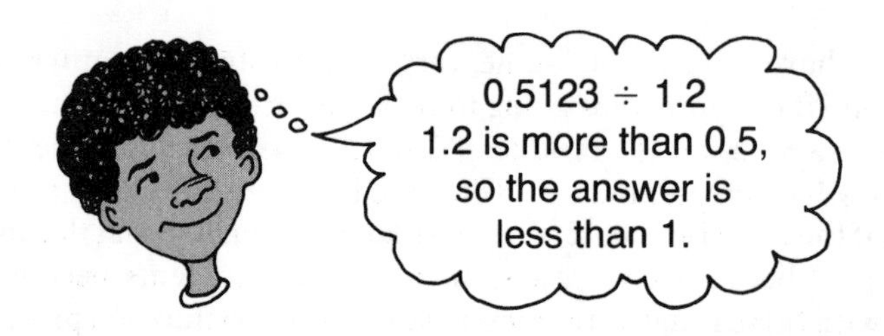

For succeeding lessons on division estimation, separate the exercises into three categories: (*a*) divisor and dividend both greater than 1, with divisor less than dividend; (*b*) divisor less than 1, with divisor less than dividend (e.g., 4.82 ÷ 0.6); and (*c*) divisor greater than dividend (e.g., 4.82 ÷ 7.3).

Both greater than 1. When both are greater than 1 and the divisor is less than the dividend, I suggest that students round both the divisor and the dividend to the nearest whole number. For example, rounding 5.286 ÷ 1.2 would give 5 ÷ 1. For closer estimates the analogy to money can again be used; both the divisor and the dividend are interpreted as money. Some of the more able students would use $5.30 ÷ $1.20 to estimate the quotient. They would probably tell me that there are more than four, but fewer than five, "piles" of $1.20 in $5.30.

Divisor less than 1; divisor less than dividend. Where the divisor is less than 1, the money interpretation is used exclusively. For 5.286 ÷ 0.12, for instance, students might use any of the following to estimate the quotient:

$5.00 ÷ $0.10
$5.30 ÷ $0.10
$5.30 ÷ $0.12

Once the rounding has been done and the results transformed into a "money problem," several different mental calculation procedures can be used. One of these procedures is to calculate multiples of the divisor. Using $5.30 ÷ $0.12, the thinking pattern might be as follows:

10 times 12 cents is $1.20 (too small).

20 times is $2.40 (too small);

40 times is $4.80 (still too small).

4 times $0.12 is $0.48, so 44 times $0.12 is $5.28.

That's close! 5.286 ÷ 0.12 is about 44.

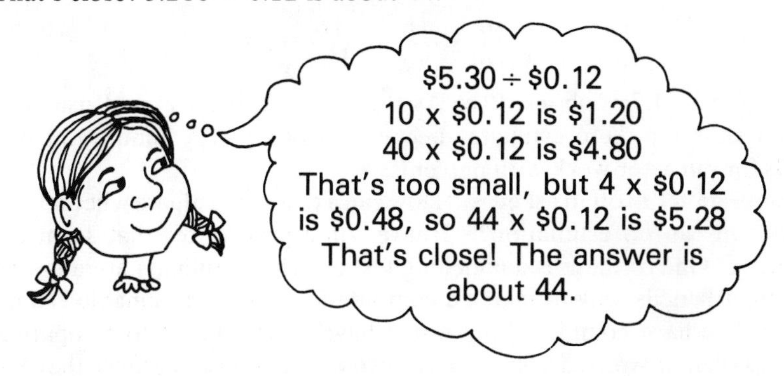

Another useful estimation strategy is to begin by using a dollar as the dividend. When we use this strategy for a problem like $5.30 ÷ 0.12, we discuss how many groups of $0.12 are in $1.00. That's a little more than 8. Five times 8 is 40, so there will be a little more than 40 groups of $0.12 in $5.00. Forty is an acceptable estimate, but we can come even closer. There are two groups of $0.12 in $0.30; 40 + 2 = 42, so the quotient will be a little more than 42.

Estimating quotients by converting a decimal problem to a money problem is effective, of course, only when the divisor and dividend are both greater than or equal to 0.005. Again I avoid examples with very small divisors and dividends, since they can be more appropriately dealt with in the study of scientific notation. However, it would be possible to use the procedures I've described above if students understand that a quotient is unchanged when both the divisor and dividend are multiplied by the same number. In an example such as 0.5286 ÷ 0.0012, the divisor and dividend might be multiplied by 100 giving 52.86 ÷ 0.12, which could then be estimated.

Divisor greater than dividend. Where the divisor is larger than the dividend, the procedure assumes that students have some knowledge of fractions and the relationship of division to fractions. For example, the problem 0.753 ÷ 4.7 is transformed to $0.80 ÷ $4.70 and then discussed in the fraction context, $0.80/$4.70. I encourage students to think, "Eighty cents is what part of $4.70?" Since 6 × $0.80 = $4.80, $0.80 is about 1/6 of $4.70. Students who have studied decimal equivalents for fractions might know that 1/6 is about 0.17. However, mine haven't, so they must also estimate the decimal for 1/6. To do this, I ask them to find a fraction for 1/6 that has a denominator close to 100, for example:

$$\frac{1 \times 15}{6 \times 15} = \frac{15}{90} \approx \frac{15}{100} = .15$$

or

$$\frac{1 \times 16}{6 \times 16} = \frac{16}{96} \approx \frac{16}{100} = .16$$

So, 0.753 ÷ 4.7 is about 0.16. As this example illustrates, I usually use hundredths for these estimates because working with hundredths builds skills for our later work with percents.

A question I'm often asked is, "How can a teacher evaluate whether or not a student can do estimation?" I have some procedures that seem to be effective. One of these is a matching test on which students are allowed to use their pencils only to match a computation with a reasonable estimate. After they have completed the test, I have them pair up to compare and discuss their answers. These teams of two do a good job because they know that when they return for a group discussion, they will be asked *why* they chose the answers they did.

Another evaluation procedure I use is to take two additional points off for every unreasonable answer on a written test. That's two points more than the number of points deducted for a wrong, but reasonable, answer. This encourages students to check the relationship between the answer and the problem. I follow this up by having students write, in complete sentences, why these answers are unreasonable. I accept no explanations such as "I copied the problem wrong," "I misplaced (or forgot) the decimal point," or "The answer is too high (or too low)." In other words, the explanation can have nothing to do with techniques. It must be about number relationships.

This program for teaching estimation works. The number of unreasonable answers decreases dramatically as the year progresses. Students can, and do, become proficient estimators.

14

Estimation and Mental Arithmetic with Percent

Glenn D. Allinger
Joseph N. Payne

OF ALL the topics we teach in mathematics, percent is one of the most useful and practical. Its use is prevalent in all phases of sports, consumer affairs, business, and government.

Considering the importance of the topic, we should obtain good results, especially for problems wih only modest computational requirements. But this is not the case. The results on local, state, and national tests are so poor as to be unacceptable, as shown by two sample questions and their results from the two most recent National Assessments of Educational Progress (NAEP 1979, p. 20; NAEP 1983, p. 15), followed by results from one of our tests (Allinger and Payne 1984, p. 17).

NAEP 1977–78

	13-year-olds Percent correct	17-year-olds Percent correct
30 is what % of 60?	35	58
What is 4% of 75?	8	27

NAEP 1981–82

	13-year-old 8th graders Percent correct
What is 10% of 50?	48
What is 60% of 50?	32

Allinger/Payne 1984 Selected pretest problems on percent for 96 eighth grade and 200 general mathematics students

Problem	8th graders Percent correct	General math students Percent correct
What % of 10 is 6?	56	40
Find 1% of 600.	45	14
Find 20% of 80.	28	8

What are the reasons for these poor results? One is the insufficient time devoted to percent in school. Both Maxim (1982) and Allinger and Payne (1984) did a survey of current textbooks with similar results. They found only about half as many pages on percent in current texts as in those from the 1950s. Apparently percent was considered more important then, and texts more nearly reflected the time needed to teach percent adequately. The sequence on percent developed by Maxim, Payne, and Allinger included substantially more lessons than most texts, and still some parts of the topics were not mastered by most students.

Another reason for poor performance is evident in interviews with students. It is obvious that the way percent has been taught encourages young people to rely exclusively on a set of rules to do all percent problems. "You do something with the decimal point. Now, which way does it go? And how many places?" Students often apply a rule blindly, moving the decimal point the wrong way or the wrong number of places. Placing a percent sign on an answer incorrectly or omitting one when it is needed are errors observed by all teachers of percent. The overriding concern of students to find and use some rule, whether chosen and performed correctly or not, shows that they have very little understanding of percent. With so little understanding, they have no choice but to try to follow some rules.

Students are not the only ones who produce glaring errors. Here are two examples from the adult world:

Example 1: A scoreboard in one of the major league baseball stadiums beamed this message:

TEAM ________ (NAME OMITTED)
BTG AV 0.247 PCT

Little did the person responsible know that this batting average is substantially below 3 hits per thousand attempts instead of the intended 247 per thousand.

Example 2: A store gave a 10 percent discount to people who had a certain discount card. To find the discount on an item, the clerk used a calculator. A customer bought a $35 item and the clerk keystroked as follows: 35 ⊗ 0.10 (%) . The result was a discount of $0.035 instead of the correct amount of $3.50. The sad part of the story is that it took six weeks for a customer to discover the error.

Both errors could easily have been avoided with the use of very elementary mental arithmetic.

The work on percent begun by Maxim and Payne (Maxim 1982) five years ago and continued by Allinger and Payne (1984) for over three years emphasizes mental arithmetic and estimation. We have found that this emphasis reduces the need to rely solely on a set of rules, since the mental arithmetic and estimation used in percent are based predominantly on understanding.

APPROACH TO PERCENT

A percent shows a comparison of two numbers, one of which is always 100. The fact that 100, a nice round number, is always employed is one of the strengths of percent. Whereas fractions and decimals often show comparisons, they seldom use 100 as the unit. A percent sign shows that the comparison is expressed as hundredths.

Most percents are *part* to *whole* comparisons. At a time when we were more precise with our language, the "part" was called "percentage" (note how often *percentage* is used by well-educated people when they actually mean *percent*), and the "whole" was called the *base.* Part and whole seem to be more easily understood by students because the words relate naturally to concrete, everyday examples. Consequently, these two words permeate our lessons.

Most approaches to teaching percent use some variation of the *proportion* method or the *factor-factor-product* method.

Proportion Method

Example 1: Paul got 3 hits out of 12 times at bat. What is his batting average as a percent?

3 is the part and 12 is the whole. Set up a proportion.

$$\frac{\text{(Part)}}{\text{(Whole)}} \qquad \frac{3}{12} = \frac{n}{100}$$

There are two main ways to solve the proportion: (1) Multiply both sides by $12 \cdot 100$ (yielding $3 \cdot 100 = n \cdot 12$); or (2) multiply by 100 (yielding $(3/12) \cdot 100 = n$). Solving gives $n = 25$. The batting average is 25 percent.

Example 2: Charles gets a 10% discount on a refrigerator priced at $700. How much is his discount?

$$\frac{\text{(Part)}}{\text{(Whole)}} \qquad \frac{n}{700} = \frac{10}{100}$$

Then $n = 700 \cdot (10/100)$ or $n = 70$, so the discount is $70.

Factor-Factor-Product Method

To use the factor-factor-product method for the same two problems, we set them up with the percent written first, the whole second, and the part third, as follows:

Example 1: ___?___ % of 12 = 3
(Whole) (Part)

$$n \cdot 12 = 3$$

Solving gives $n = 3/12$, or $1/4$. In other words, $n = 0.25$, which is expressed as 25%.

Example 2: 10% of 700 = n
(Whole) (Part)

$$0.10 \cdot 700 = n, \text{ so } 70 = n$$

Note that the process involves writing the percent, usually as a decimal (it could be a fraction) and multiplying.

Maxim (1982) investigated the two approaches to percent with seventh-grade students and found no major differences in achievement results. He did note a slight tendency in favor of the factor-factor-product method. However, there is hardly a topic about which teachers have such strong views concerning their preferred method of teaching as percent. In informal surveys, we have found that slightly more than half of the middle school/junior high school teachers prefer the proportion method. It seems safe to say that a teacher who has had success with one method can satisfactorily continue with that method. Teachers who use the proportion method must eventually teach the correct keystroking sequence for the calculator—that is, teach students to think of 7 percent of 800 as 800 · 7% rather than 800 · (7/100).

In the initial study by Maxim (1982), mental arithmetic and estimation were important components of the sequence. In subsequent studies by Allinger in general mathematics classes, even greater emphasis was given to mental arithmetic and estimation, and the proportion method was used exclusively.

MENTAL ARITHMETIC

By *mental arithmetic* we mean the process of finding exact answers mentally without the aid of paper and pencil or calculator. By *estimation* we mean finding a number that is in the neighborhood of the exact answer, with the size of the neighborhood depending on the purpose. For example, a student who is doing poorly and gets 36 out of 85 test items correct may consider an estimate of 40 percent close enough for scoring, whereas one who gets 79 out of 85 right may wish to estimate her score to the nearest tenth of a percent (92.9%), especially if 93 percent is the cut-off point between grades A and B. This highlights one of the difficulties with estimation—deciding how close an estimate is needed. It is difficult to make rigid rules.

Approximate numbers may be involved in the mental arithmetic process, that is, some numbers may be replaced with "easier" numbers. In estimating 50% of 27, it could be useful to think of 27 as 30 or 20 or 28. For each, approximation has been used to transform 27 into another number, the first by rounding to the nearest ten, the second by front-end truncation, and the third by adjusting digits to a more compatible number (50% = 1/2 and 1/2 of 28 is 14).

Mental arithmetic is almost always an important part of estimation, so mental arithmetic comes first. In our percent lessons, we begin the mental arithmetic sequence with easy unit fractions. We use these fractions to assist with other fractions, giving special attention to thirds and eighths, and then go on to find 1%, 10%, 50%, and 100% of a number.

Easy Unit Fractions and Their Percent Equivalents

We begin with the meaning of percent as hundredths. We relate the meaning to fractions because of the ease of associating fractions with part-to-whole comparisons. We use equivalent fractions to generate hundredths for the easy fractions, the fractions whose denominators are factors of 100.

Meaning of percent: $\frac{5}{100}$ = 5 hundredths = 5%

$\frac{83}{100}$ = 83 hundredths = 83%

Using equivalent fractions:

$$\frac{1}{2} = \frac{50}{100} = 50\% \qquad \frac{1}{4} = \frac{25}{100} = 25\% \qquad \frac{1}{5} = \frac{20}{100} = 20\%$$

$$\frac{1}{10} = \frac{10}{100} = 10\% \qquad \frac{1}{25} = \frac{4}{100} = 4\% \qquad \frac{1}{50} = \frac{2}{100} = 2\%$$

The fractions are shown two ways, using diagrams with 100 squares to represent hundredths (fig. 14.1). The diagrams reinforce the well-understood concept of a part-to-whole comparison.

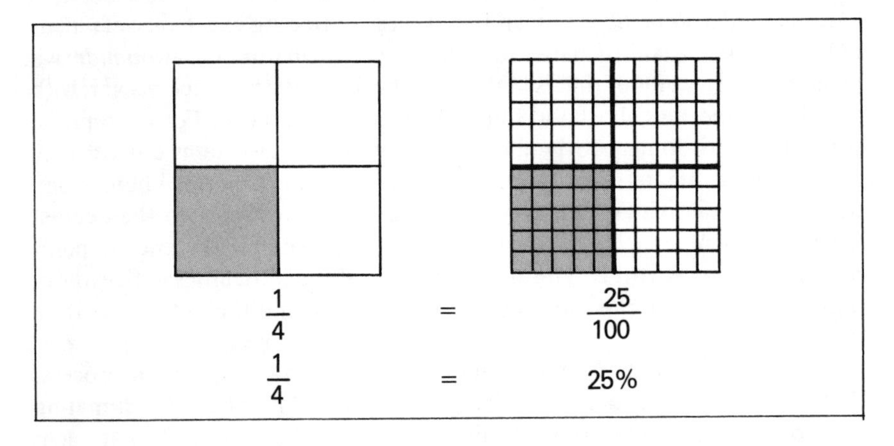

Fig. 14.1

Students memorize the easy unit fraction–percent equivalents, being expected to give the percent for a fraction or the fraction for a percent in about two seconds. The unit fractions are then used to derive the percents for fractions with the same denominator.

A	B
Since $\frac{1}{4} = 25\%$,	Since $\frac{1}{5} = 20\%$,
$\frac{3}{4} = 3 \times 25\%$,	$\frac{4}{5} = 4 \times 20\%$,
so $\frac{3}{4} = 75\%$	so $\frac{4}{5} = 80\%$

Fraction-Decimal-Percent Equivalents for Easy Fractions

The place of decimals in a percent sequence can vary greatly. In his sequence for general mathematics, Allinger (in press) did most of the work with percent using fractions and proportions. A subsequent revision by Allinger & Payne included decimals as the last four lessons. Allinger found that introducing decimals early caused added confusion for students. With the exclusive use of proportions for percent, there is little real need for decimals. In a factor-factor-product approach, decimals must come early because, for example, 27% of 863 is found by multiplying 0.27 and 863.

Since comparisons can be represented by fractions, decimals, or percents, a well-rounded curriculum must contain all three. Our goal should be for students to attain flexibility in moving from any one to the other two. For example, to find a 25% discount mentally, you would find 1/4 of the original price; for a batting average of 0.330, you might think of 1/3, or about 30%.

We expect students to demonstrate this flexibility initially by giving the two missing numbers for the easy unit fractions in figure 14.2 in about four seconds.

Fraction	Decimal	Percent
$\frac{1}{2}$	?	?
?	0.25	?
?	?	10%

Fig. 14.2

Thirds and Eighths

Maxim (1982) analyzed the strengths and weaknesses of the cognitive structures of students. He found that they have much greater difficulty with thirds and eighths than with the easy fractions. For thirds, the dilemma is that the decimal and percent do not "come out even." For eighths, the difficulty seemed to be the need for an additional decimal place for thousandths. For both, it is obviously necessary to determine whether the decimal is to be rounded to hundredths and the percent to a whole number. For thirds, there is the additional need to tell students when an exact answer is required and when to approximate with a rounded number. For example,

$$\frac{1}{3} = 0.3\overline{3} = 3\overline{3}\% \quad \text{(exact)}$$

$$\frac{1}{3} \approx 0.33 = 33\% \quad \text{(rounded to hundredths and nearest whole percent)}$$

$$\frac{1}{3} \approx 0.333 = 33.3\% \quad \text{(rounded to thousandths and tenth of a percent)}$$

For most paper-and-pencil calculations, and for all on the calculator, the decimal approximation is used. In a similar way, the decimal representation for 2/3 is truncated, yielding 0.66 (66%), or rounded to 0.67 (67%). One-eighth is treated exactly as 0.125 (12.5%) or rounded to 0.13 (13%).

Students are expected to learn decimal-percent equivalents for 1/3 and 1/8 and then derive them for 2/3 and also the "eighths"—2/8 through 7/8.

In summary, the initial mental arithmetic for fraction-percent-decimal equivalents stresses a part-to-whole comparison, the use of the hundred square as a visual model, and the basic meaning of percent as hundredths. The aim is for students to develop flexible thought in moving from one form to the other two with reasonable speed and accuracy.

Finding 1%, 10%, 50%, and 100% of a Number

The easier percents—1, 10, 50, and 100—are used to find exact answers, with some care exercised in the initial choice of base numbers. They are also employed to estimate answers for practical purposes and as a check on calculations.

One hundred percent is easiest to understand when the concept of percent is expressed using models, since 100% = 100/100 = 1, or "everything"—for instance, 100% of 3749 is 3749.

To find 50% of a number, 1/2 is usually used.

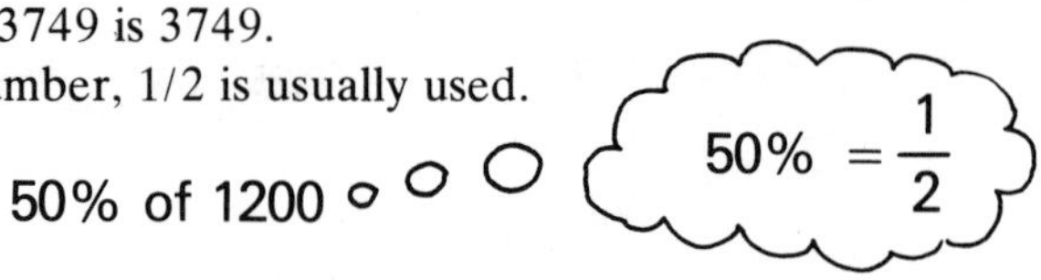

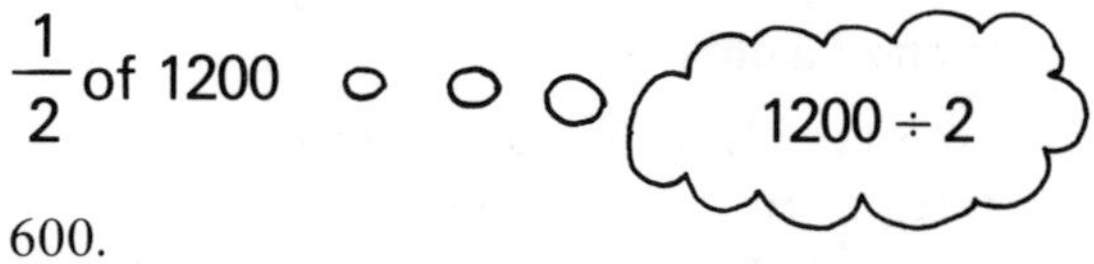

50% of 1200 ○ ○ ○ (thought: $50\% = \frac{1}{2}$)

$\frac{1}{2}$ of 1200 ○ ○ ○ (thought: $1200 \div 2$)

And so, 50% of 1200 = 600.

One percent of a number can be found by proportions or by the factor-factor-product method. For 1% of 1600, for instance, the two methods are illustrated below.

A	**B**
$\frac{1}{100} = \frac{n}{1600}$	$1\% \text{ of } 1600 = n$
$(1/100) \cdot 1600 = n$	$0.01 \cdot 1600 = n$
$16 = n$	$16 = n$

Note that the decimal point is moved two places to the left. (The fact that 1600 is the same as 1600. [decimal point added] needs to be pointed out to students.)

Similarly, to calculate 10% of a number, we use 10/100 = 1/10, or move the decimal point one place to the left.

10% of 860 ○ ○ ○ (thought: 860. / 86.0 / 86)

A mental algorithm for dealing with 10% and 1% creates the quandary of whether to move the decimal one or two places to the left. Posing the following question can be helpful: Which is larger, 10% of a number or 1% of that number? Referencing the hundred-square model or the fact that $10\% = 10 \cdot 1\%$ answers the query.

This knowledge of how to calculate 1% and 10% is now used to mentally calculate other percents.

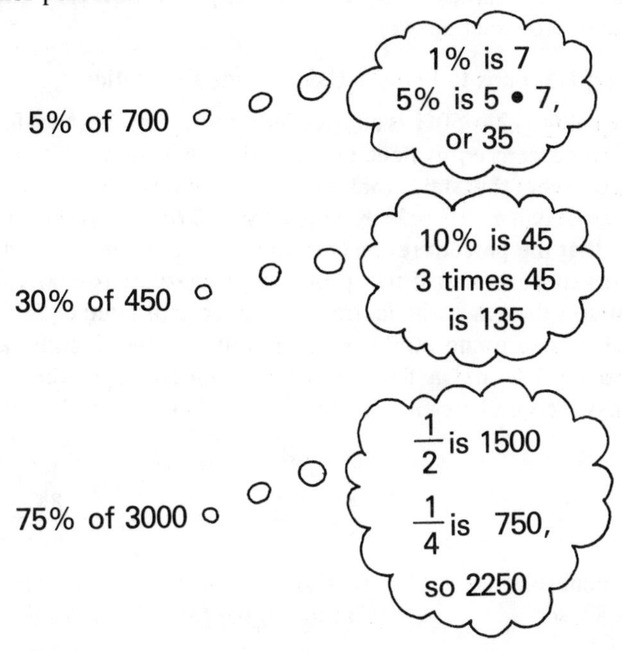

Although these examples use numbers that are easy to work with mentally, this mental arithmetic stresses the meaning of percent, fractions, and decimals. It will become evident in the following section that mental arithmetic is a highly useful tool for estimating answers in practical problems and computation.

ESTIMATION

To estimate means to find an answer in the neighborhood of the exact answer. To be practical, estimation must be fast and easy, accentuating the need for mental arithmetic. Estimation could possibly involve the use of paper and pencil or calculator, depending on the complexity of the problem, the type of numbers involved, and the technique used. For example, estimating the number of people in an auditorium could be done mentally, but estimating the cost of remodeling the classroom might be more involved. Our work with percent emphasizes estimation with mental arithmetic.

An estimate may sometimes answer a question directly. Will \$5.00 cover the purchase of an item priced at \$6.10 less a 30 percent discount? The customer who can calculate mentally can proceed confidently to the check-out stand. At other times, the estimate will provide a check on the exact answer, a skill required by the clerk who handles your money. In any event, the validity of the estimation should be derived from the mathematics concepts and techniques used, not from comparing how close the estimate is to the supposed exact answer.

Converting Fractions to Percent Using Crude Estimation

The equality $1/2 = 50\%$ is an excellent, though rough, tool for estimating fraction-to-percent equivalencies. A student will struggle to estimate percent in a problem that states that 8 of 17 classmates are present. Now, $1/2 = 8/17$ if and only if $1 \cdot 17 = 2 \cdot 8$, so $8/17 \neq 1/2$ (or 50%); but the inequality indicates that the percent representation for 8/17 is in the neighborhood of 50%. This suggests that if two times the numerator is nearly equal to the denominator, then the original fraction can be estimated by 50%. Similarly, a fraction whose numerator is small relative to the denominator may be estimated by 1% and a fraction whose numerator is very close to the denominator could be estimated by 100%. Three examples follow.

A	B	C
$\frac{2}{87} = \frac{?}{}\%$	$\frac{41}{87} = \frac{?}{}\%$	$\frac{84}{87} = \frac{?}{}\%$
2 is small relative to 87, so	$2 \cdot 41 = 82$, which is close to 87, so	84 is almost 87, so
$\frac{2}{87} \approx 1\%$	$\frac{41}{87} \approx 50\%$	$\frac{84}{87} \approx 100\%$

Converting Fractions to Percent with Refined Estimates

Closer estimates can be made by comparing the fraction to 1/10, 1/4, and 3/4, but the difficulty level rises.

Example 1

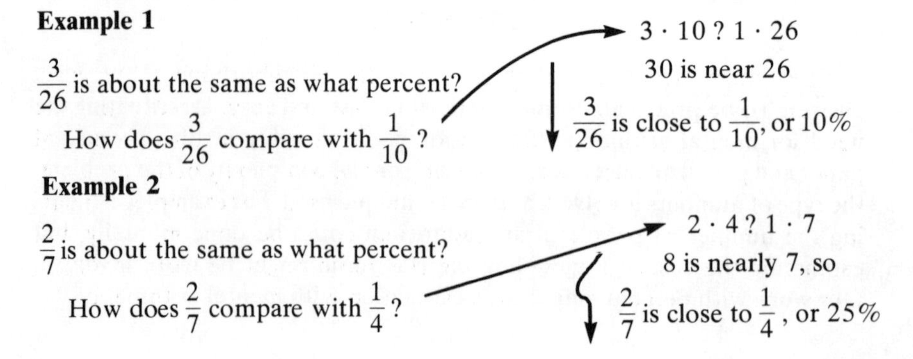

$\frac{3}{26}$ is about the same as what percent?
How does $\frac{3}{26}$ compare with $\frac{1}{10}$?
$3 \cdot 10 ? 1 \cdot 26$
30 is near 26
$\frac{3}{26}$ is close to $\frac{1}{10}$, or 10%
Example 2
$\frac{2}{7}$ is about the same as what percent?
How does $\frac{2}{7}$ compare with $\frac{1}{4}$?
$2 \cdot 4 ? 1 \cdot 7$
8 is nearly 7, so
$\frac{2}{7}$ is close to $\frac{1}{4}$, or 25%

Students who can handle this technique should be challenged by the question: Is the estimation greater or less than the exact answer?

Converting Fractions to Percent: Other Techniques

Another method of estimating the percent equivalent of a given fraction involves writing fractions with a denominator of 100 that are equivalent to the given fraction. Although computing 7/20 = 35/100 = 35% is relatively easy, this is not true of 6/19. But 19 · 5 is nearly 100, so an equivalent fraction is 30/95 or about 30/100, or 30%.

Example: If 7 of 9 cones are chocolate, what percent are chocolate?

$$\frac{7}{9} = \frac{?}{100}$$

What number times 9 is close to 100?

$$9 \cdot 11 = 99$$

$$\frac{7}{9} \cdot \frac{11}{11} = \frac{77}{99}$$

$$\frac{77}{99} \approx \frac{77}{100} = 77\%$$

A variation of the method above would involve replacing the given denominator with a more compatible number. For example, 7/9 is easier to deal with as 7/10, which is 70/100 = 70%. Likewise, 6/19 is easier when converted to 6/20, since 20 is a more compatible number than 19: 6/20 = 30/100 = 30%. Of course, the compatible numbers to be used as denominators are the factors of 100 (e.g., 1, 2, 4, 5, 10, 20, 50, 100). The question of whether estimates are higher or lower than the exact percent depends on an understanding of fractions—in particular, the fact that replacing the denominator with a larger (smaller) integer decreases (increases) the rational number.

Finding a Percent Using "EZ%"

We created the term EZ% (read "easy percent") for 1%, 10%, 50%, and 100%. Instead of computing 13% of 800, say, students can estimate by substituting 10%—the closest EZ% to 13%: "10% of 800 is what?" We have found that one lesson devoted simply to transforming "hard" exercises into a related EZ% exercise is time well spent. Initially, exact answers to the original exercises can be ignored. Some class time must be directed toward determining which EZ% is nearest the given percent, paying particular attention to 30% and 75%. Rounding, truncating, and compatible numbers play a significant role also.

Example: What is 61% of 847?

Original problem may be transformed in one of these ways:

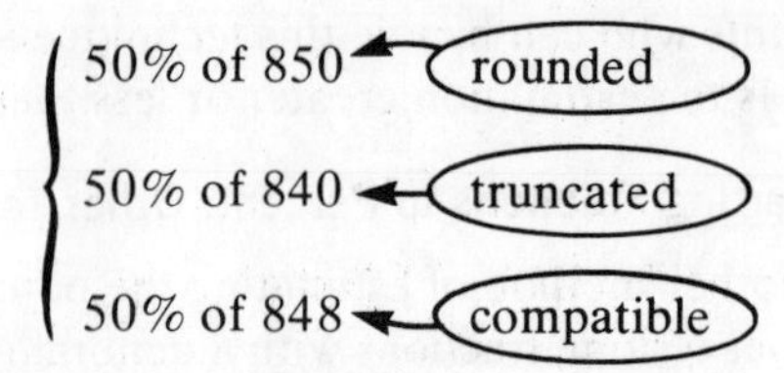

A refinement would involve replacing the given percent with a much closer approximation. For example, approximate 61% by 60% where 60% = 6 · 10%. So the mental estimation process would be:

Original exercise

Other Percents of a Number

Since 30% and 75% are exactly betweeen two EZ% values, they can be used for still more accurate estimates. Thus it is helpful to know their fractional equivalents: 30% is nearly 33%, or approximately 1/3, and 75% = 3/4. Other helpful equivalents, mentioned previously in the section on mental arithmetic, are 25% = 1/4, 67% ≈ 2/3, 12.5% = 1/8. The problem, "What is 15% of 18?" can be replaced by, "What is 12.5% of 16?" yielding (1/8) · 16 = 2 as a reasonable estimate.

Adjusting or Bounding Answers

To adjust an estimate is to increase (or decrease) it so that it is closer to, or in the neighborhood of, the exact answer. This is a higher-level skill than simply estimating. Consider this problem:

> Judy Resnik, a member of the *Discovery* crew, worked 57% of the 140 hours she spent in space. How many hours did she work?

EZ% problem:	50% of 140 = 70
Adjustment:	Add 10% more to the answer, or 14, so 70 + 14 = 84.
Adjust again:	Subtract about 4 more, since we are still high, so 84 − 4 = 80.

Flexibility in the estimation process is important.

The amount of adjustment can also be determined by using bounds on the exact answer, that is, overestimates and underestimates. When computing 8% of 427, for example, determine an underestimate and overestimate as follows:

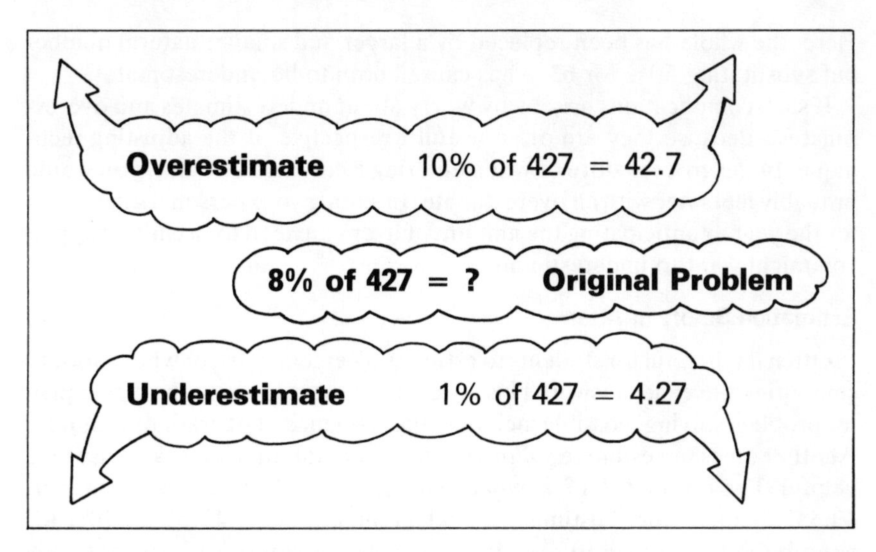

The exact answer is somewhere between 4 and 43, but since 8% is closer to 10%, an adjusted estimate could be 35. We have found that writing this information as

$$1\% \text{ of } 427 < 8\% \text{ of } 427 < 10\% \text{ of } 427$$

causes panic in an eighth-grade classroom. Although correct, such a mathematical statement seems to camouflage the issue.

Overestimating and underestimating are likely to raise more complex questions than a single adjustment. For 63% of 287, a student may suggest using the bounding technique in the following *erroneous* manner:

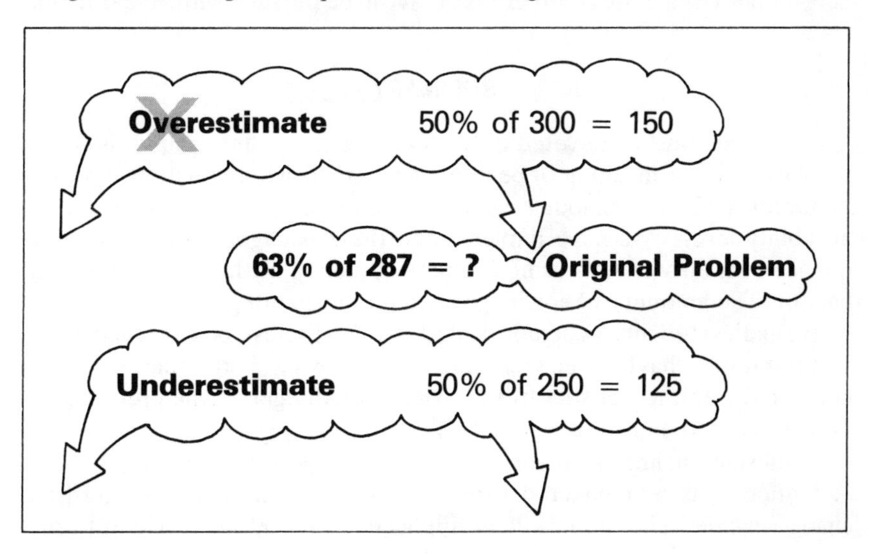

Here, the whole has been replaced by a larger and smaller natural number but substituting 50% for 63% has caused both to be underestimates.

If such complications arise, why worry about underestimates and overestimates? Because they are often useful irrespective of the adjusting technique. In the grocery store or when entering a crowded elevator, you would probably feel safer with an overestimate. In setting your personal sales quota for the year or anticipating the amount of liver you need for a child's supper, you might want to underestimate.

Estimation Before or After

Often it is helpful for students to estimate a percent, part, or whole prior to computing the exact answer. This encourages them to think through a plan for problem solving, possibly including the sequence of operations required. At other times, an estimate *following* the computation of an exact answer is helpful. For instance, "17 is what percent of 40?" has an exact answer, 42.5%. For an "after" estimate, a student might think, "42.5% of 40 leads to an EZ% statement: 50% of 40 = 20, which is close to 17. Occasionally using both "before" and "after" estimates can build added confidence in an answer.

In this section, we have seen that fast and easy estimation is aided by truncating or rounding numbers or replacing them with compatible numbers in preparation for mental arithmetic. Recognizing relationships between numerators and denominators can be valuable in estimating fractional equivalents for percent. Learning to use EZ%—namely, 1%, 10%, 50%, and 100%—provides the first step for transforming a problem. Refining this process by adjusting estimates, including the use of overestimates and underestimates, is a more complex task that can be pursued with able students.

SUMMARY

More time must be devoted to the teaching of percent. Emphasis should be placed on the meaning of percent as a part-to-whole relationship using hundredths. Concrete models can be effectively employed with students of all ability levels. We have implemented these pedagogical ideas in a sequence of twenty lessons using estimation, mental arithmetic, and the calculator. The amount of knowledge acquired by eighth graders and general mathematics students, as demonstrated through pretest, posttest, and retention test results, has been encouraging. The following posttest data should be compared with the statistics from NAEP and Allinger/Payne pretest questions listed in the introduction to this article.

A substantial improvement can be seen in performance at all levels. Retention tests administered three weeks later indicated no significant change in comprehension, skill, or application. Yet in the course of twenty

Allinger/Payne: 1984	Selected posttest problems on percent for 98 eighth graders and 199 general mathematics students	
Problem	8th graders Percent correct	general math students Percent correct
What % of 10 is 6?	93	69
30 is what % of 60?	91	73
What % of 300 is 9?	*89 (83)	*54 (72)
Find 1% of 630.	97	66
Find 12% of 60.	87	42
Find 4% of 75.	*90 (96)	*53 (84)

* Indicates question was given twice, once on Part I of the posttest without calculators and repeated on Part II with calculators available (% in parentheses).

lessons there was only time enough to deal with two of the three types of percent problems. Also percents less than 1 and greater than 100 were not included for mastery. Clearly, the teaching of percent requires more time than it receives in current textbooks.

Mental arithmetic must be practiced daily, particularly computations dealing with converting "hard" fractions (thirds and eighths) to percent. Estimation should be used not only as a check on exact answers but to provide meaning for concepts of fraction, decimal, and percent. Always demand estimation, but not necessarily prior to the completion of the problem. Let the estimation process justify its worth independent of how close the estimate is to the exact answer. Use the calculator as a constructive instrument to help develop concepts and to compute. Minimize its use as a tool for estimation, and instead promote mental arithmetic as the fast and easy process.

REFERENCES

Allinger, Glenn D. "Percent, Calculators, and General Mathematics." *School Science and Mathematics,* in press.

Allinger, Glenn D., and Joseph N. Payne. "Teaching Percent to General Mathematics Students." Unpublished manuscript. Department of Mathematical Sciences, Montana State University, Bozeman, 1984.

Maxim, Bruce R. "An Evaluation of Two Sequences for Percent." Ph.D. dissertation, University of Michigan, 1982.

National Assessment of Educational Progress. *Mathematical Knowledge and Skills—Selected Results from the Second Assessment of Mathematics.* Denver: Education Commission of the States, 1979.

———. *The Third National Mathematics Assessment: Results, Trends, and Issues.* Denver: Education Commission of the States, 1983.

Benchmark: Measuring Mount Everest

Barbara J. Reys

A SUPPOSEDLY true story told by a friend illustrates how the appearance of numbers affects the way we interpret them. It involves an effort to measure the height of Mount Everest some years ago. A team of surveyors was commissioned to measure the height of the mountain for official use in reference books. Since all measurement involves some error, the crew decided on a plan to make six independent measurements, then average them for their "official" measure. The six measurements were collected and averaged:

	28 990
	28 991
	28 994
	28 998
	29 001
	29 026
Total:	174 000
Average:	174 000 ÷ 6 = 29 000 feet

When the average was calculated to be 29 000 feet, the crew was concerned. They feared that readers would view 29 000 feet (with all those zeros) as a much less accurate measure than was actually the case. They finally decided to change the figure to 29 002 feet so that its appearance would reflect the careful measurement techniques from which it was derived. For many years 29 002 feet was reported as the official height of Mount Everest!

15

Varieties of Estimation

Rheta N. Rubenstein

ESTIMATION opportunities are plentiful and diverse:

- If pens cost $2.29 each, about how much will a dozen cost?
- In what place will the quotient begin when 5497 is divided by 27?
- About how much will it cost to retile our roof?
- About how many schoolchildren are there in the United States?
- About how many schoolchildren will there be in the United States in 2006?
- How long will it take us to drive from Chicago to Toronto?
- Is $50 going to be enough to pay today's grocery bill?
- My calculator says 786 × 9 equals 6184. Does this seem reasonable?
- What is the least and the most a student can expect to pay for a freshman year in college?

Although most estimation situations share the common feature of finding an approximate answer to a problem in a brief period of time (Reys et al. 1980), they differ from each other in many ways. Some involve measuring, some calculating, some surveying, some predicting. They differ in complexity, in context, in the amount of information required, in numbers involved, in presentation, in the type of answer expected.

There are a number of reasons why educators should be aware of the varieties of estimation situations and the relationships among them. First of all, seeing the varieties helps us appreciate the pervasiveness of estimation in our lives and the significance it deserves in the mathematics curriculum. Sharing this appreciation with students should be an important affective goal of the curriculum. Also, recognizing the distinctions, similarities, and relationships increases one's understanding of what estimation is.

Second, in designing and implementing curricula with estimation objectives, educators should be aware of the scope of possibilities in order to provide systematic development and sequencing of materials. For example, teachers need to make decisions based on questions like these: Which types of estimation items are simpler? What sequence will promote the development of estimation skills? What estimation instruction can be used in conjunction with instruction in problem solving? How do certain estimation situations match youngsters' psychological development? What skills are related to estimation performance (e.g., distinguishing rounded from exact numbers, operating with rounded numbers, identifying orders of magnitude, using scientific notation) and how should instruction and practice with them be woven into estimation instruction? To answer these questions and to develop creative, efficient instruction that incorporates estimation into lessons that have broader goals require a full awareness of the varieties of estimation.

Third, in evaluating students' performance in estimation, teaches should be aware of the variety of possible measures and be able to match appropriate measures to objectives in the curriculum.

Finally, in designing and interpreting research in estimation, educators need to have a common vocabulary and comparable measurement instruments derived from a survey of the possibilities.

Thus, it is important to be aware of the variety in estimation situations. This article explores this variety from several perspectives and discusses some curricular implications related to each.

VARIETIES IN LEVEL OF DIFFICULTY

Like other situations in mathematics, an estimation situation may be an exercise or a problem. If the student knows immediately how to proceed, the situation is an exercise; otherwise, it is a problem. For example, for an upper elementary grade student, the following items should be exercises:

- Will the quotient when 875 is divided by 12 be closer to 7, 70, or 700?
- My calculator says 786 + 276 = 952. Does this seem reasonable?
- If oranges cost $1 a dozen, about how much does one cost?

The following item, however, should be a problem for almost everyone:

- About how many households will there be in the United States twenty years from now?

The intent of much mathematics instruction is to turn problems into exercises by teaching appropriate skills and concepts. To that end, estimation instruction can bolster the learning of number concepts and processes. For example, items like the first three listed above can be used not only to develop estimation skills but also to promote the learning of place value, order of magnitude, number facts, and meaning of operations.

The more difficult estimation problems (and even some of the "exercises") share some process characteristics with problem solving. Both use processes like "use simpler numbers," "choose the appropriate operation," and "look back" to see if the answer makes sense. Providing instruction on general processes such as these may improve both estimation and problem solving.

VARIETIES IN THE AVAILABILITY OF DATA

Estimation situations may come with all the necessary data for solution, or they may require some measurements or research. When measurements are required, there are a number of possibilties. First, the measure needed could be discrete (how many?) or continuous (how much?). Estimating discrete measures can be facilitated by sampling, ratio and proportion, cross sections, and other techniques as suggested, for example, in *Peas and Particles* (Elementary Science Study 1966). Bright (1976) has neatly cataloged the varieties of situations requiring estimation of continuous measures: the unit of measure may or may not be present, the object to be measured may or may not be present, or the measure may be given with the student required to find an object of that measure.

When research is required, it may be accessible (e.g., to estimate the grocery bill, list the costs of the items in the cart) or inaccessible. Consider the following:

- About how many schoolchildren are there in the United States? Here the data required are probably inaccessible. Furthermore, depending on the availability of time and resources and the teacher's objective, this could be a social studies problem involving finding references, subdividing the country, and so forth. More pertinent to mathematics classes, accessible data such as the number of students in one's own school district and the size of the community might be used to develop a ratio that could then be applied to the U.S. population.

VARIETIES IN TYPES OF NUMBERS

Estimation opportunities exist with *all* types of numbers, not merely with the commonly used whole and decimal numbers. Middle and high school students should be able to use estimation in ways such as those shown in figure 15.1.

VARIETIES IN THE FORMAT OF THE QUESTION

Estimation questions may be presented in a variety of formats. In an *open-ended* format (e.g., If pens cost $2.29 each, about how much will a dozen cost?), no guidelines are given for the answer. This is the most common style of "real life" estimation. *Reasonable vs. unreasonable* estimation (e.g., My calculator says 786 multiplied by 9 equals 6184. Does this

Fig. 15.1

seem reasonable?) asks the student to check the reasonableness of a "calculated" answer by comparing it to an estimate. This skill is becoming more and more important with the widespread use of, and dependence on, calculators. *Reference number* estimation (e.g., Is $50 going to be enough to pay today's grocery bill?) asks the student to decide whether the answer to an estimation item is over or under a given reference number. *Interval* estimation (e.g., What is the least and the most a student can expect to pay for a freshman year in college?) asks the student to identify upper and lower bounds for an answer. *Multiple choice* estimation asks the student to choose from given answers the one closest to the exact answer.

This last format is used on most standardized tests. Although multiple choice items are generally not useful in evaluating estimation performance (Reys and Bestgen 1981), they may have a role in developing estimation skills if they are judiciously designed. For example, one type of multiple choice question asks students to focus on *orders of magnitude* (e.g., Is 12×28 closest to 30, 300, or 3000?).

Research has shown that student performance varies significantly depending on which format is used (Rubenstein 1983, 1985). Instruction that uses the easier formats (order of magnitude or reference number) before expecting students to do more difficult types (open-ended) should promote more successful learning.

VARIETIES IN THE STYLE OF PRESENTATION

Opportunities for estimation can be presented to youngsters in several ways. Nearly all verbal exercises in mathematics textbooks at the prealgebra level can be presented as estimation items. These items include many exercises of the puzzle type as well as conventional problems. Teachers can present estimation items orally or on the overhead projector with short timed intervals for students' responses. Also, teachers can use "mini quizzes"—sets of estimation items that are presented on small sheets of paper with no scratch space and that must be completed in a brief time.

VARIETIES IN THE TYPES OF ANSWERS

Estimation situations differ in the types of answers that are meaningful. Exactness is not always possible or preferred. Consider questions like these:

- About how tall are eleven-year-olds?
- How much will it cost to heat our home next winter?

Although estimates are often inexact answers arrived at because conditions such as a lack of time or information do not allow exact calculation, many times (as in the examples above) an estimate is the *only* possible or meaningful response. Generally, these situations are categorized in the realm of statistics. The best answer may be the "most likely" value or a "confidence interval." Exposing youngsters to these sorts of estimating situations beginning in the middle grades can not only prepare them to appreciate statistics but also can help extinguish the idea that mathematics always seeks exact answers.

SUMMARY

Varieties of differences among estimation situations have been discussed. It is hoped that this discussion will help teachers, curriculum developers, textbook authors, and researchers create successful estimation learning experiences for youngsters.

REFERENCES

Bright, George W. "Estimation as Part of Learning to Measure." In *Measurement in School Mathematics,* 1976 Yearbook of the National Council of Teachers of Mathematics, edited by Doyal Nelson, pp. 87–104. Reston, Va.: The Council, 1976.

Elementary Science Study. *Peas and Particles.* Newton, Mass.: Educational Development Center, 1966.

Reys, Robert E., and Barbara J. Bestgen. "Teaching and Assessing Computational Estimation Skills." *Elementary School Journal* 82 (November 1981): 117–27.

Reys, Robert E., Barbara J. Bestgen, James F. Rybolt, and J. Wendell Wyatt. *Identification and Characterization of Computational Estimation Processes Used by Inschool Pupils and Out-of-School Adults.* Washington, D.C.: National Institute of Education, 1980.

Rubenstein, Rheta N. "Mathematical Variables Related to Computational Estimation." *Dissertation Abstracts International* 44 (1983): 695A. (University Microfilms No. 83-06935)

———. "Computational Estimation and Related Mathematical Skills." *Journal for Research in Mathematics Education* 16 (March 1985): 106–19.

16

A Do-It-Yourself Estimation Workshop

Julia A. King

THE activities described in this article are drawn from a workshop for the Elementary Mathematics Curriculum Committee of the School District of Philadelphia, a group of administrators, supervisors, and teachers. The activities were selected to illustrate a variety of uses of estimation as well as important techniques and skills involved in estimating accurately. An effort was made to choose examples that would appeal to adults and also be suitable for use in the classroom. Activities that had not previously been tried were tested in the classroom and evaluated by the staff before being included in the workshop.

It quickly became apparent in the planning process that estimating enters into decisions made in dozens of daily activities; this makes it a potential link between mathematics and nearly every other part of the curriculum. Suggestions for activities came from cooking, sewing, shopping, drawing, woodworking, energy use, driving, travel, sports, business, home finance, child rearing, politics, history, and many other spheres of endeavor. Skillful estimating, it appeared, could save one from being late, being cheated, spoiling a snapshot, overpaying income tax, and a host of other calamities. The examples chosen for the workshop represent only a few of the hundreds of possibilities.

A large laminated poster (fig. 16.1), designed for classroom use, greeted the participants as they entered, creating a lively interest in estimating.

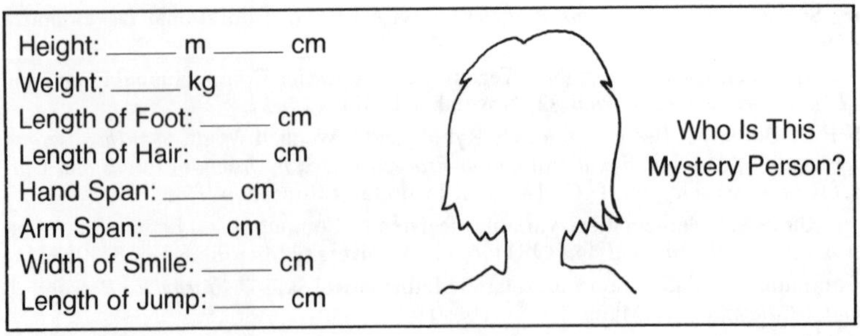

Height: _____ m _____ cm
Weight: _____ kg
Length of Foot: _____ cm
Length of Hair: _____ cm
Hand Span: _____ cm
Arm Span: _____ cm
Width of Smile: _____ cm
Length of Jump: _____ cm

Who Is This Mystery Person?

Fig. 16.1

One workshop participant had kindly volunteered her measurements, and they were entered on the poster ahead of time. The poster's effect on the adults was the same as its effect had been on children when it was tested. It aroused curiosity instantly. Participants began to eye each other, speculating about personal statistics. Estimates abounded. It was promised that the identity of the mystery person would be revealed at the end of the workshop.

Twenty other activities were set up as learning stations. Participants were invited to choose partners with whom to discuss and try the activities. When all had visited each station, estimates and techniques were discussed and everyone computed his or her score, using a scoring system based on what the staff considered reasonable results.

THE ACTIVITIES

Instruction cards for many of the activities, each listing the equipment supplied, are presented here, followed by comments on techniques for estimating, modifications for classroom use, possible extensions, and other observations. Appropriate grade levels are indicated.

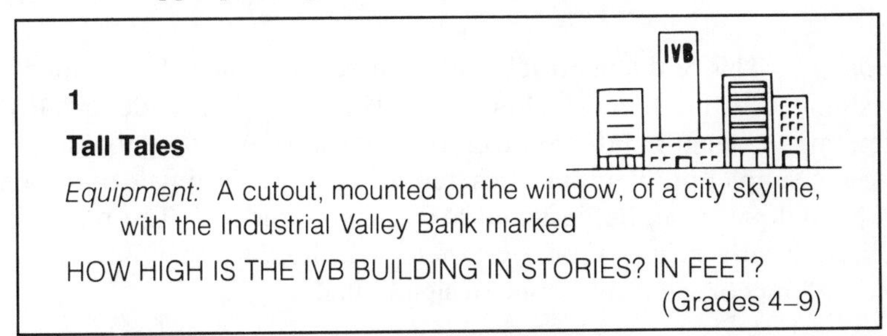

1

Tall Tales

Equipment: A cutout, mounted on the window, of a city skyline, with the Industrial Valley Bank marked

HOW HIGH IS THE IVB BUILDING IN STORIES? IN FEET?

(Grades 4–9)

Comments: Only the upper portion of the building was visible, so its height in stories had to be estimated by comparing it with other buildings that could be seen in their entirety. An estimate of the height in feet can be based on a reasonable estimate of the height of a story. Of course, not all buildings have stories the same height. Measuring and comparing the height of ceilings in various types of buildings (old and new houses, schools, and office buildings) to establish a rule of thumb could be an interesting individual or group project.

2

Hand Spans

Equipment: A centimeter ruler taped to a table

MEASURE YOUR HAND SPAN IN CENTIMETERS. USE YOUR HAND SPAN TO ESTIMATE THE LENGTH OF THE TABLE IN CENTIMETERS.

(Grades 3–8)

Comments: Personal measurements are probably the most useful of all tools in estimating lengths and distances, since they are always at hand. Furthermore, children delight in measuring their own hand spans, arm spans, paces, and nose-to-fingertip reaches and find these easy to remember. This activity is one of hundreds that make use of personal measurements in estimation.

3

Interstate Travel

Equipment: A map, including a scale of miles, with a route traced in ink; string or pipe cleaners

HOW LONG WILL IT TAKE TO DRIVE FROM HAGERSTOWN, MARYLAND, TO HUNTINGTON, WEST VIRGINIA (TO THE NEAREST QUARTER OF AN HOUR) AT AN AVERAGE SPEED OF 50 MPH?

(Grades 8–12)

Comments: This problem must be solved in several stages. "How long" can be estimated if one knows "How far." "How far" can be determined by estimating the length of the route on the map and converting to miles using the scale. Simpler problems like this, requiring only an estimate of the length of a curved path, can be designed for younger students. The path can be duplicated with string or pipe cleaners, which can either be marked in standard units or measured after straightening.

4

The Sheriff and the Dinosaur

Equipment: A life-sized tracing of a child (in sheriff's dress) and a small drawing of a triceratops

THE NOSE HORN OF THIS TRICERATOPS IS AS LONG AS ONE FOUR-YEAR-OLD SHERIFF. ESTIMATE HOW MANY SHERIFFS LONG THE TRICERATOPS IS.

(Grades K–3)

Comments: Even four-year-olds can estimate. Estimation goes hand-in-hand with learning the meaning of a unit of measure. The technique of using one's own height for comparison is useful in later life as well. One can estimate the height of a tree, for example, by visualizing one's own height marked off along its trunk.

5

Plan Ahead

Equipment: Letter stencils, taped to the table; cardboard

ESTIMATE HOW MANY LETTERS OF THE ALPHABET (STARTING WITH **A**) WILL FIT ACROSS THIS PIECE OF CARDBOARD. NO OVERLAPPING!

(Grades 1–10)

Comments: How often we fail to make such an estimate before beginning to letter! The popularity of PLAN AHEAD signs attests to this. An estimate can be made using finger widths or other informal units of measure. It should be noted, however, that different letters occupy different amounts of space, and that in order to space lettering well, one must take into account what letters are being used.

6

The Black Spot

Equipment: A template for drawing circles and a card with a circular black spot

ESTIMATE WHICH CIRCLE ON THE TEMPLATE WAS USED TO DRAW THIS SPOT.

(Grades 2–8)

7

Nuts and Bolts

Equipment: Three bolts (identified with letters) and six nuts (identified with numbers) mounted on a board so they cannot be moved

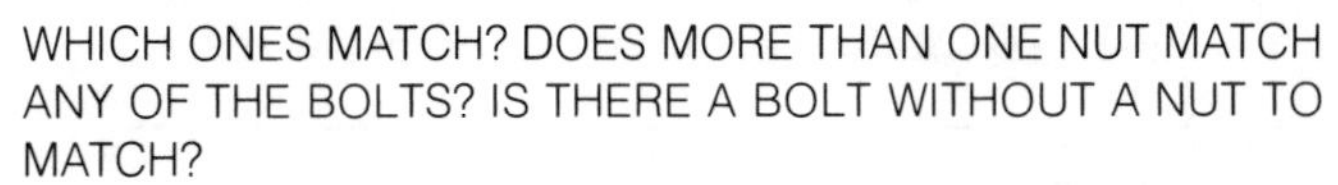

WHICH ONES MATCH? DOES MORE THAN ONE NUT MATCH ANY OF THE BOLTS? IS THERE A BOLT WITHOUT A NUT TO MATCH?

(Grades 4–8)

Comments: Preschool children solve a problem similar to this one when they fit a set of graduated cylinders into matching holes. The cylinder task is simpler than this, however, since the one-to-one correspondence between pegs and holes makes it necessary only to arrange the pegs in serial order to determine which peg fits which hole. (Of course, most preschoolers do not solve the problem in this way.) This problem is also more difficult because the nuts and bolts are stationary.

8

Quilting Quotient

Equipment: A color photograph of a quilt; a timer

YOU HAVE ONE MINUTE TO EXAMINE THIS QUILT BEFORE ESTIMATING WHAT PERCENT OF IT IS WHITE. (Grades 7–10)

Comments: Estimating percent is a skill often used in daily life. Activities such as this one not only help students to develop that skill but also may serve as introductory exercises, enabling students to gain an intuitive understanding of percentage before dealing with it formally. One valuable technique in making an estimate like the one required here is to ask, "Is it more than 50 percent?" "Less than 75 percent?" and so forth.

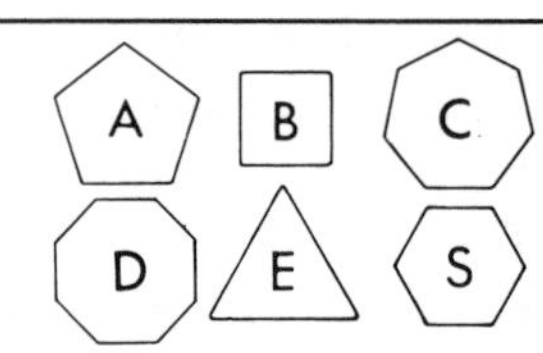

9

Shapely Relations

Equipment: Six cutout polygonal regions labeled A, B, C, D, E, and S

COMPARE THE AREAS OF SHAPES A, B, C, D, AND E WITH THE AREA OF SHAPE S. DECIDE WHETHER EACH ONE IS GREATER THAN, EQUAL TO, OR SMALLER THAN S.

(Grades 7–10)

Comments: One might attempt to make a numerical appraisal of the area of each polygon, or superimpose one on another for comparison, or mentally subdivide each into regions of approximately equal size. This type of problem lends itself well to presentation on an overhead projector.

10

Cook's Eye

Equipment: A jug of water, pan, measuring cup, and a sponge to wipe up any spills

POUR WATER FROM THE JUG INTO THE PAN, STOPPING WHEN YOU ESTIMATE YOU HAVE POURED A CUPFUL. USE THE MEASURING CUP TO CHECK HOW MUCH YOU POURED.

(Grades 3–7)

Comments: The aim of this activity is to develop the ability to judge a standard unit of liquid measure. Estimates should improve if the activity is repeated, giving one the opportunity to become familiar with the containers and to compare estimates with those made previously. In the classroom, many types of containers other than a pan can be used: mugs, teacups, bottles and jars, a thermos, a fish bowl, a bucket. Pints, quarts, gallons, or liters may be substituted for cups. A teacher may want to invite children to guess the level to which a standard unit will fill a container, then demonstrate by pouring.

11

Just Another Drop

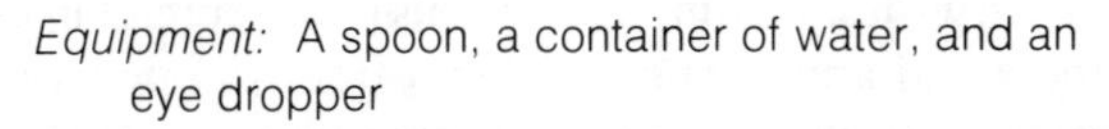

Equipment: A spoon, a container of water, and an eye dropper

FILL YOUR SPOON WITH WATER. FILL THE EYE DROPPER WITH WATER. ESTIMATE HOW MANY DROPS OF WATER YOU WILL BE ABLE TO ADD TO THE SPOONFUL BEFORE THE WATER IN THE SPOON OVERFLOWS. (Grades K–8)

Comments: Often the ability to make good estimates depends not so much on skill in calculating as on a familiarity with physical phenomena—here, the effects of surface tension. Many people are astonished by the results of this experiment. Again, estimates should improve rapidly as the procedure is repeated.

12

A Loaded Question

Equipment: A packed suitcase

HOW HEAVY IS THIS SUITCASE IN KILOGRAMS? (Grades 5–12)

Comments: Estimating weight is difficult enough, even in the English units to which most of us are accustomed, because we perceive small, dense objects as "heavier" than large, less dense things. (Remember the old riddle, "Which is heavier, a pound of nails or a pound of feathers?") The only effective technique seems to be comparison with other objects of similar density whose weights we recollect. Therefore it is helpful to be familiar with the weight and "heft" of a number of commonplace objects that can serve as standards for comparison: a nickel (5 g), a two-liter bottle of soda (2 kg), a quart of milk (2 lb.), a ten-pound bag of potatoes, and so on. Such familiar objects may be included for comparison in this activity.

13

Is a Picture Worth a Thousand Words?

Equipment: An article from the newspaper of well over a thousand words; five pictures of different sizes and shapes

WHICH PICTURE IS WORTH A THOUSAND WORDS? THAT IS, WHICH ONE IS NEAREST IN SIZE TO 1000 WORDS PRINTED IN THIS SIZE TYPE.

(Grades 4–9)

Comments: Estimating the number of words needed to fill a space is a practical problem for the writer or compositor. This is an exercise in estimating both large numbers and areas. The problem offers an opportunity for introducing older students to the column inch as a unit of measure.

14

Thick or Thin

Equipment: A ruler marked in centimeters and a stack of paper at least 1 cm thick

ESTIMATE THE THICKNESS OF A SHEET OF THIS PAPER IN MILLIMETERS.

(Grades 6–12)

Comments: No specialized instruments are needed for this estimate. Counting the number of sheets in a one-centimeter stack and dividing by ten will give the number of sheets on a one-millimeter pile. One more division will produce the desired estimate. Estimates of the size of small objects can often be made by measuring the objects in aggregate and dividing by the number of objects.

15

Longest Light in Town

Equipment: A traffic light, visible from the room

ESTIMATE HOW MANY SECONDS THE TRAFFIC LIGHT ON THE CORNER REMAINS RED. DO NOT USE A TIMING DEVICE.

(Grades 3–6)

Comments: Many people have learned to estimate time by counting in such a way that each count is about one second. Another strategy is counting heartbeats or breaths, once a normal rate has been established.

16

Checkout

Equipment: A shopping cart or bag filled with grocery items; a timer

HOW MUCH WILL IT COST YOU TO CHECK OUT?

Time Limit: One minute

Caution: Underestimates score 0.

(Grades 6–12)

Comments: Not having enough money is embarrassing at the cash register; hence the 0 score for underestimates. Shoppers could doubtless suggest dozens of methods for making estimates. Possibilities include assigning an average price to all items (say, $1.00), grouping items totaling approximately a whole number of dollars, or rounding each price to the nearest dime or quarter.

17

Rooting Around

Equipment: Paper and pencil; a calculator with no square root function; a timer

TAKE THREE MINUTES TO ESTIMATE $\sqrt{8464}$ USING THIS CALCULATOR.

(Grades 8–12)

Comments: Making arithmetical estimates has been widely discussed and is perhaps the type of estimation most familiar to teachers. This problem was selected as an illustration for adults both because it is thought-provoking and because most adults have forgotten any algorithm for extracting square roots. Estimation, then, is the only way to arrive at a solution.

Estimating square roots not only is a practical skill but also serves to fix in mind the meaning of "square root" itself. Even when a calculator is available, it is helpful to know the squares of multiples of 10. Since the square of 90 is 8100 and the square of 100 is 10 000, the square root of 8464 can quickly be bracketed between 90 and 100. The calculator can then be used to check the squares of numbers in between. As it happens, the answer is a whole number.

It is easy to simplify this problem by choosing a much smaller perfect square, or to make it more difficult by selecting a large number whose square root is irrational and requiring an estimate correct to the nearest tenth or hundredth.

18

Middle Age

WHAT IS THE MEDIAN AGE OF THE PEOPLE AT THIS WORKSHOP?

(Adults)

Comments: This last problem is just for fun. Participants were asked in advance to submit their ages (anonymously) on slips of paper so that the actual median age could be calculated.

CONCLUSIONS

A workshop such as this can be organized by any group wishing to explore the topic or share its own ideas and experiences with others. Only two staff members of the Philadelphia Teacher Parent Center, which presented this workshop, were mathematics educators. The others specialized in language arts, early childhood education, the arts, and the design of educational equipment. Nevertheless, all contributed equally to the planning and preparation.

What made this joint effort possible was an initial brainstorming session, the purpose of which was simply to get ideas on the table without any judgments or analysis. Since nearly everyone uses estimation consciously or unconsciously every day, all could contribute. The mix of suggestions was, in fact, much richer than it would have been had the planning been restricted to those with experience in teaching mathematics.

Such a procedure allows a group—whether teachers, administrators, or staff developers—to design a workshop tailored to the interests, experience, and comfort level of its own members. A brainstorming session, followed by a work period in which ideas are sorted and evaluated and specific activities are designed, can constitute an entire workshop for a group wishing to explore estimation on its own. Brainstorming alone can be a stimulating introductory activity in a workshop conducted for others, since it creates an awareness of the enormous range of situations in which estimating is useful and exposes the techniques we already know and apply.

17

"Guess and Tell": An Estimation Game

Beth M. Schlesinger

THE mathematics classroom is a place for learning—on the part of both the student and the teacher. One day, when discussing a trigonometry test, I asked the class: What would be a good estimate for 10/0.6428?

Expecting a resounding chorus of "fifteen," I was shocked to hear dead silence and then a few mumbled answers, such as 60, 0.6, and even 0.06. I then realized that my bright and mathematically talented students could not successfully estimate the answer because they had never had much practice in estimation. That night the estimation game "Guess and Tell" was born.

"Guess and Tell" is a game that requires students to make estimations and mental calculations. It's an oral, class participation game that deals with serious (and some not-so-serious) material in a spontaneous and nonthreatening way. Its main goal is to get students to feel more at home in the physical world of measurements and the mathematical world of abstractions.

MANAGEMENT AND SCORING

To play "Guess and Tell," the class is first divided into groups of equal size and similar mathematical ability. Three nonplaying roles are needed: problem reader, scorekeeper, and judge. These roles can be combined, but the teacher should serve as judge, at least at first. One group is chosen at random to get the first problem. A problem is read to the first student in this group.

Without pencil or paper or any calculating device, the student gives an estimate of the answer. The judge decides if the estimate is close enough to the exact answer. If it is, the side is awarded five points. If the student will not estimate, the other side is allowed to make an estimate. If their answer is close enough, they will get the five points. Since guessing is not penalized, it is rare that a student will not guess. At first students may be reluctant to make an estimation, but once the game gets under way, their hesitancy will vanish. When this game is managed well, it can be highly enjoyable for both students and teacher. At the end of the game, the group with the most points is the winner.

Again, the game should be kept on the light side, even though its purpose is very serious. It can be played for short or long periods of time. Sample questions for precalculus students follow; these can be modified for any level from upper elementary through college. The teacher must be careful to gear the problems to the level of the class involved. The teacher will want to keep a collection of problems with answers, possibly on index cards. The deck of cards can be shuffled and problems selected at random. Students should also be encouraged to contribute problems—with solutions—and they should be given credit on the problem card.

SAMPLE PROBLEMS

The problems I use for this game are of three basic types: review, factual, and whimsical. If an overhead projector is available, small transparencies of visual problems could be prepared.

Problem	Approximate Answer and Rationale
1. $\sqrt{78} \approx ?$	8.8. Answer is between 8 and 9 and is much closer to 9.
2. $3.1 \times 18.6 \approx ?$	57 or 58. Answer is between 54 and 60.
3. Which is larger: 10^3 or 3^{10}?	3^{10}. $10^3 = 1000$, and since $2^{10} = 1024$, 3^{10} is certainly larger.
4. $\log_3 35 \approx ?$	3.2. Answer is between 3 and 4 and is closer to 3.
5. Estimate the solution to this set of equations: $x + y = 10$, $y = 2x$	A quick substitution gives $3x = 10$, so $x \approx 3.3$ and $y \approx 6.6$.

6. What is the approximate measure of each interior angle of a regular 14-gon?	As the number of sides of a regular polygon increases, the number of degrees in each interior angle also increases. For a regular hexagon the angle is 120° and for a regular octagon the angle is 135°; so 150° would be a good guess. The exact answer is 154.3°.
7. How many black "pentagons" are on a soccer ball?	This answer requires visualization. There are about five on the top hemisphere and five on the bottom. Ten would be a good guess. There are actually 12.
8. Which has a larger area: a square of side 2 or a circle with radius 1?	The area of the square = $2^2 = 4$. The area of the circle is approximately $1 \times 1 \times 3.14 \approx 3.14$. So the square is larger.
9. What is the approximate value of tan 70°?	A good estimate is 2. Tan 60° = $\sqrt{3} \approx 1.732$. So tan 70° ≈ 2.
10. What is sec x if x is in Quadrant II?	Cosine is negative in Quadrant II, and so is the secant. The secant will have to be any value less than −1. That is all that can be said. This question is purposely vague to generate discussion.

FACTUAL PROBLEMS

Writing these problems will involve some research in standard reference works, such as encyclopedias, almanacs, and record books. They can be serious or fun. The main objective is to get the students to lose their inhibitions and guess.

Problem	Answer
1. How long is the Great Wall of China?	1500 miles
2. What is the population of your city—in my case, San Diego?	About 800 000
3. What is the world's record for the number of hard-boiled eggs consumed in one minute?	14 (by Peter Dowdeswell of Bristol, England, in 1978; source: *1982 Gui-*

ness Book of World Records)

WHIMSICAL PROBLEMS

The first of these problems came from the September 1983 issue of *Games Magazine*. The second, inspired by it, is one of mine.

1. You have a loud voice. In fact, your voice is so loud that when you yell "hello" from New York City, a friend in Los Angeles can hear you. After you yell "hello," how long does it take before your friend hears your voice?

   Choose one: Seconds, Minutes, Hours, Days, Weeks, Months, Years, Decades, Centuries

   Answer: 2600 miles $\times$ 1 hour/742 miles $\approx$ 3.5 hours

2. John has straight hair of average length. One morning he wakes up and finds that all his hair has fallen out! To console himself, he places the hairs end to end to see how far they will reach. What is the closest estimate?

   1 hm 1 km 10 km 100 km 1000 km

   Answer: 110 000 hairs $\times$ 0.1 m/1 hair = 11 km
   So 10 km is the closest estimate.

JUDGING THE ANSWERS

Skill is needed to judge the game as well as to play it. The judge must decide whether the estimate is close enough to earn points, "close, but no cigar," or "out in left field."

For the problem $10/0.6428 \approx ?$, a calculator yields 15.56. I would accept any estimate in the range of 14–18. Other teachers might prefer a wider or narrower range of estimates.

Now and then you will have students who are such strong estimators that it will be possible for them to serve as judges, providing the class will accept their authority.

CONCLUSION

This game will encourage students to estimate and, moreover, it will give them confidence in their guesses and mental calculations. Perhaps it will also make estimation more of a habit, one that they will use throughout their work in mathematics and science as well as in their day-to-day lives.

18

Fermi Problems

or How to Make the Most of What You Already Know

Joan Ross
Marc Ross

THE famous physicist Enrico Fermi (1901–1954) frequently amused, and sometimes startled, his listeners by posing problems. One favorite Fermi problem was "How many piano tuners are there in Chicago?" The essence of a Fermi problem is that a well-informed person can solve it (approximately) by a series of estimates. It also has the characteristic that most people who encounter it instantly respond that this is a problem that they could not possibly solve without recourse to some reference material. Another characteristic is that most people can be led to a solution through a series of simple steps that use only common sense and numbers that are either generally known or amenable to estimation by recollections of direct observation. Fermi used such problems—and many other physicists continue to do so—to make an important educational point: problem-solving ability is often limited *not* by incomplete information but by the inability to use the information that is already available. Many physicists and computer scientists seem to be adept at solving Fermi problems. Perhaps it is now time for mathematics teachers and their students to become more aware of Fermi problems and some ways to go about solving them.

In this article we shall analyze two Fermi problems. In each instance we shall pose the problem—at which point perhaps the reader would like to try to come up with a method and an estimate. Then we shall give two different formulas for solving each problem, which will allow us to make independent estimates. (Again the reader is invited to try to come up with appropriate values for the formula pieces; remember, it all has to come from generally known information.) Finally, we shall give our approximate numbers, the methods by which we arrived at them, and our final results.

We would like to thank Fred Hendel, Roy Meyers, Art Schwartz, and Beth Spencer for their helpful suggestions.

THE POPCORN PROBLEM

This problem was suggested by a cautionary tale told to us by Pamèla Ames, then a teacher in the Lab School of the University of Chicago. Her class was working on metric measurements, and somehow she found herself, her class, and eventually the whole school hard at work creating a cubic meter of popcorn. It took them more than a week. Here's the related Fermi problem:

How many kernels of popcorn are there in a cubic meter of popped corn?

Formula 1

Number of kernels = volume in cubic centimeters of a cubic meter/volume in cc of a piece of popped corn
= 10^6/ ("diameter" in cm of a piece of popped corn)^3.

Note that since the dimension of the popcorn piece is to be cubed, it needs to be more accurately estimated than quantities that appear to lower powers.

Formula 2

Number of kernels = number of kernels in 2 tablespoons of unpopped corn × (1 cubic meter/1 liter).

To understand this formula, you need to know (as this article's popcorn-loving authors did) that it takes 2 tablespoons of unpopped kernels to make a liter of popcorn.

Estimations

1. We take the diameter of a piece of popped corn to be 1.5 cm. Then the number of kernels $\approx 10^6/3.5 \approx 3 \times 10^5$.
2. We make an estimate of 120 kernels of corn per tablespoon (see below). (A direct measurement of this quantity is easy to make but not exactly in the spirit of a Fermi problem. We do not wish to discourage direct measurement, but we do wish to emphasize the Fermi process.) This leads to the result that

$$\text{the number of kernels} \approx 240 \times 10^3 \approx 2 \times 10^5.$$

You may say that you could not have estimated the number of kernels in a tablespoon. But here's an approach to use when you're especially uncertain: Estimate lower and upper limits and take the geometric mean. In this example, what can you say about the least number of kernels in a tablespoon? Picture a tablespoon measure and put kernels of corn into it. You know that many more than ten kernels will fit in the spoon. What about twenty? Let us say that you are unsure whether more than thirty kernels will fit. Using the same technique, let us say you are unsure whether fewer than

500 kernels will fit. The geometric mean is the square root of the product of the two numbers:

$$\sqrt{30 \times 500} \approx 120$$

(This happens to be an excellent estimate.) It is surprising but true that knowing that a number lies between 30 and 500 is powerful information. The reason one uses the geometric mean is that it minimizes the maximum percentage error.

INTERLUDE

One of us was trained in mathematics, the other in physics. The mathematician sympathizes with readers who are made queasy by the *grossness* of these reckonings. In many mathematical contexts, we might be distressed by getting an answer of 3×10^5 by one method and an answer of 2×10^5 by another. Nevertheless, in the Fermi-problem context, we are delighted by how well these two answers agree. And isn't that what estimation is all about? There are so many times in the real world when all you really want is an order-of-magnitude estimate. For example, newspapers are full of numbers; these numbers are frequently misstated by a factor of a thousand (e.g., the newspaper states a million when it should have stated a billion). An experienced Fermi-problem solver has the advantage of being able to detect quickly that some numbers are as reasonable as others are ridiculous.

Notice also how much conventional mathematics creeps into the problems. Area and volume and percentage just keep cropping up. Another characteristic worth noting is that we are constantly dealing with units and unit conversions. Sometimes the metric system is more appropriate, and sometimes we believe the English system is more useful (because our basic information uses both systems).

The next problem is much more complex and difficult than the first. A natural response may be that "I (or my students) could never do that!" Be assured that you (and they) can. But it does take practice and a willingness to be very "dirty" (of which, more later).

THE FOREST/DOLLAR PROBLEM

One of us recently attended a Michigan Council of Teachers of Mathematics meeting in which someone pointed out that the U.S. budget has become a topic that demands the use of scientific notation. A listener responded that she had been told that if all the trees in the U.S. were cut down and the wood made into dollar bills, the total value would not add up to the 1984 U.S. budget.

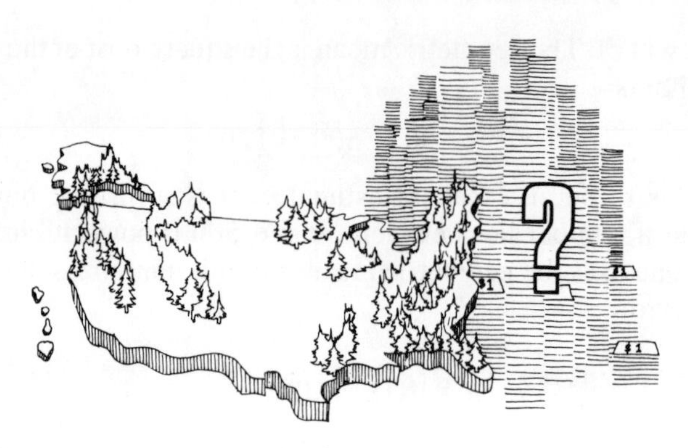

Fermi problem: How many dollar bills would you get from all the trees in the U.S.?

Underlying Formula

Number of dollars = number of pounds of wood in U.S. trees × number of dollar bills per pound of wood.

We propose these two ways to find the number of pounds of wood:

Formula 1

Number of pounds of wood = number of U.S. trees × number of pounds of wood per tree.

Formula 2

Number of pounds of wood = number of pounds of wood used yearly to make paper in the U.S./fraction of total standing wood used to make paper each year.

Estimations

This more complex problem shows the recursive nature of solving Fermi problems. Most of the quantities in our two original formulas now become Fermi subproblems. Of course, we always have in mind the characteristic of the formulas that we must eventually achieve: each quantity involved must be amenable to estimation.

We shall deal first with the number of dollar bills per pound of wood, estimating (here and below) that a pound of paper is made from a pound of wood. Now we think of some paper whose weight we know. For example, we know that in mailing letters we can send about 5 standard sheets at the basic rate. Thus, 5 sheets of paper weigh about 1 ounce (the maximum weight permitted at the basic rate). With this in mind, we use the following formula:

Number of dollar bills in 1 pound = number of bills in 1 sheet of paper × number of sheets of paper in 1 pound.

The first number is about 6 (think of cutting a sheet of paper into dollar bill shapes), and the second was estimated above to be 16×5. So the number of dollar bills in a pound $\approx 6 \times 16 \times 5 \approx 500$. (Of course, this quantity can also be found by experiment. We would not expect to get exactly the same answer.)

Now let's go on to the number of pounds of wood in U.S. trees, first using formula 1, then formula 2.

Formula 1

The parts we need to estimate are the number of U.S. trees and the number of pounds of wood per tree.

To get the number of U.S. trees, we use this formula:

$$\text{The number of U.S. trees} \\ = \text{area of U.S. forests} \times \text{number of trees per unit area.}$$

We estimate the area of U.S. forests as follows: The area of the U.S. is about 1000 miles × 3000 miles. The fraction of the U.S. that is forested surely lies between 0.05 and 0.5. We take the geometric mean (0.15) of these two numbers as our estimate. Thus,

$$\begin{aligned}\text{the area of the U.S. that is forested} &\approx 3 \times 10^6 \times 0.15 \text{ square miles} \\ &\approx 5 \times 10^5 \text{ square miles.}\end{aligned}$$

The number of trees per unit area is approximately the reciprocal of the square of the average distance between trees. We take that average distance to be 30 feet (the approximate geometric mean of 100 feet and 10 feet). Thus,

$$\begin{aligned}\text{the number of trees per unit area} &\approx 1/(30)^2 \text{ per square foot.} \\ &= \left(\frac{5280}{30}\right)^2 \text{ per square mile} \\ &\approx 3 \times 10^4 \text{ per square mile.}\end{aligned}$$

So, the number of U.S. trees $\approx 3 \times 10^4 \times 5 \times 10^5 = 1.5 \times 10^{10}$.

Now we have only to estimate the number of pounds of wood in a tree. We estimate that trees average 50 feet in height and are 1 foot in diameter.

Thus,

$$\text{the volume of an average tree} \approx \left(\frac{\pi \times 50}{4}\right) \text{ feet}^3.$$

The weight of a cubic foot of wood can be estimated by recalling that wood is just a bit lighter than water. One of us knew that a cubic foot of water weighs about 60 pounds (the other would have had to work through unit changes from knowing "a pint's a pound," etc.). So one cubic foot of wood weighs about 50 pounds. Thus,

$$\text{the weight of wood per tree} \approx \left(\frac{\pi \times 50}{4}\right) \times 50 \text{ pounds} \approx 2 \times 10^3 \text{ pounds.}$$

Finally, putting all this information into the underlying formula, we get the following:

$$\begin{aligned}\text{number of dollars in U.S. trees} &= \text{number of trees} \times \text{weight per tree} \\ &\quad \times \text{dollars per unit weight} \\ &\approx (1.5 \times 10^{10}) \times (2 \times 10^3) \times 500 \\ &= 1.5 \times 10^{16}.\end{aligned}$$

The 1984 U.S. budget is about 8×10^9 dollars, so things are not quite as bad as some of us feared.

Formula 2

We need to estimate the number of pounds of paper used yearly in the U.S. and the fraction of total wood used yearly for paper in the U.S.

We estimate the first quantity by multiplying the use per person by the population. Direct observation leads us to believe that we each throw away about 1 pound of paper a day. (In a household of three, the dry weight of paper in the trash can each week would then be 21 pounds. Does that sound about right?) Then,

$$\begin{aligned}\text{number of pounds of paper used yearly in the U.S.} &\approx 365 \times 230 \text{ million} \\ &\approx 8 \times 10^{10}.\end{aligned}$$

The fraction of wood used yearly for paper is equal to the product of the fraction of all wood that is harvested yearly and the fraction of the harvest that is used for paper. We estimate the fraction of wood harvested yearly as one-half the yearly fractional growth (the fraction of forest that grows in a year). The yearly fractional growth is about 1/(average age of a tree at harvest). This average age we take to be 30 years (the approximate geometric mean of 20 and 50). The fraction harvested for paper (as opposed to lumber or firewood), we take to be 1/2. Then,

$$\text{the fraction of wood used yearly for paper} \approx \frac{1}{2} \times \frac{1}{30} \times \frac{1}{2} = \frac{1}{120}.$$

Putting all this information into the underlying formula, we get the following:

$$\begin{aligned}\text{number of dollars in U.S. trees} &= (\text{number of pounds of paper used} \\ &\quad \text{yearly/fraction of total wood used} \\ &\quad \text{for paper}) \times \text{number of dollar bills in} \\ &\quad \text{a pound of wood} \\ &\approx (8 \times 10^{10}/(1/120)) \times 500 \\ &\approx 5 \times 10^{15}.\end{aligned}$$

The answer is 1/3 of the first answer, but that doesn't disturb us. It's not a bad agreement given the far-flung estimates required—and it appears to be good enough to confirm our first answer to the question about the U.S. budget. We can be reasonably certain that the roughness of our estimates has not led us to an incorrect conclusion.

DISCUSSION

We frequently see students who are terrified of just plunging into a new problem—especially if it looks strange and they don't know the algorithm (as if there always were one!). Too often in mathematics the emphasis has been on processes guaranteed to produce exactitude. Solving Fermi problems is a good way to loosen up by getting some good, if dirty, results.

To start estimating, one needs only to venture a number based on some quantitative connection, a number that is more than a wild guess. It is often true that we have rough ideas of upper and lower limits even when we don't "know" a number. The experienced estimator may quickly think of more than one path to obtain a quantity and may have an idea of the accuracy of the estimates, but that isn't necessary to get started.

Not every quantitative estimation problem in the real world is a Fermi problem. The distinguishing characteristic of a Fermi problem is a total reliance on information that is stored away in the head of the problem solver. Many of the nicest Fermi problems are such that their answers could *not* be looked up. (Our enthusiasm for Fermi problems should not be taken as implying that we disapprove of looking things up, asking informed people, or making fresh observations.) Solving Fermi problems presents an artificial challenge that should be fun—but it also provides a good opportunity to use many techniques that are useful in solving all kinds of quantitative problems.

We close with some more examples of Fermi problems, hoping to encourage you to solve a few on your own, to invent some others, and to introduce some into your classroom.

- How many railroad freight cars are there in the U.S.? (This problem was posed by Fermi himself.)
- How many times is a number stated by a baseball announcer in the course of broadcasting a game?
- How many thirteen-year-olds are there in the U.S.?
- How many words does a typical teenager read in a week?
- How many gas stations are there in your county?
- How many musical groups are there in the U.S. that play for money?
- How many gallons of gasoline does a large airliner consume per mile?

We would like to make a collection of Fermi problems suited to high school students. If you dream up or encounter a good one, please send it to us at the University of Michigan, Ann Arbor, MI 48109.

19

A Calculator Estimation Activity

Grayson H. Wheatley
James Hersberger

AN ACTIVITY that can be used throughout the year and that is applicable to many lessons is a valuable instructional tool for busy teachers. This article describes an exceptionally flexible, extendable estimation activity and presents a rationale for its use. The activity, "The Range Game," can be easily modified for use at many grade levels and with many mathematical topics.

"THE RANGE GAME"

"The Range Game" builds estimation skills. The only tools needed are a calculator and paper and pencil for record keeping. The object is to find numbers within a given range that will satisfy a given equation. Introduce the game by displaying a range and a partial equation, as shown in figure 19.1.

Example:	**Range**
	50
15 + ______ = ______	
	40

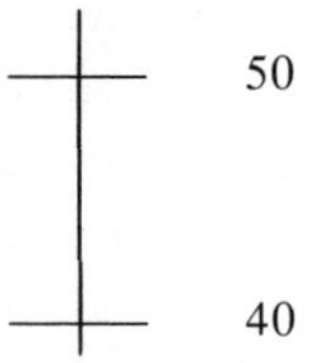

Fig. 19.1

Ask questions such as the following:

- "Find a number that when added to 15 gives a sum in the range shown at the right."
- "What is the largest number that works?" "The smallest?"

- "Are there any other numbers?"
- "Let's find all the numbers that will work."

List responses on the board and discuss the findings. Ask, "How many numbers work?"

A calculator is particularly useful in playing the Range Game. You can have students enter the number shown, the operation, and their proposed number. In the problem shown in figure 19.1, a student might try 10 and find that 15 + 10 is too small, out of the range. Because of this finding, she might then wish to try a larger number. If students are familiar with the constant addend built into most calculators, additional trials can be made by just entering a new number and pressing equals. Once one value satisfying the conditions is obtained, students tend to look for another. Finally, the complete solution set can be listed. It is interesting that students are often motivated to find all solutions even though only one was initially sought.

The activity can be used with the whole class, in small groups, or individually. Initially, introducing the Range Game to the whole class allows the teacher to clarify the rules and the students to generate excitement by interactions. Students working in small groups can tackle quite challenging variations of the Range Game. Finally, individuals can work on a set of Range Game problems individually, in class or as homework.

As the Range Game is played, the question of domain invariably arises. What numbers can be used? For example, in the problem shown in figure 19.1, if one student says there are eleven numbers in the solution set, another is likely to ask, "What about 25.5? It is in the range." This is an excellent time to discuss the concept of domain; there is a context for the concept, and meaning can be easily established. At times you may wish to specify the set of numbers to be used: whole numbers, integers, rational numbers, or real numbers.

Many other problems can be created by varying the number, operation, and range. A few examples appear in figure 19.2.

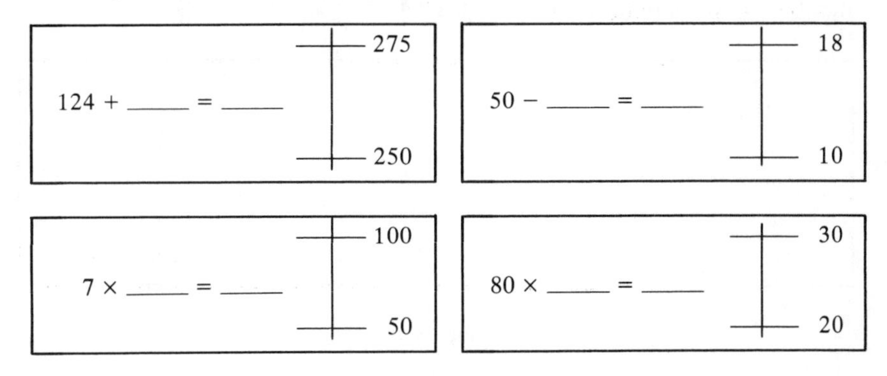

Fig. 19.2

The level of challenge can be increased by progressively making the range smaller or by choosing larger numbers. Another useful variation is to have students compare the results from problems that use different operations (see fig. 19.3).

25 + _____ = _____ (range 127 to 77)

Problem A

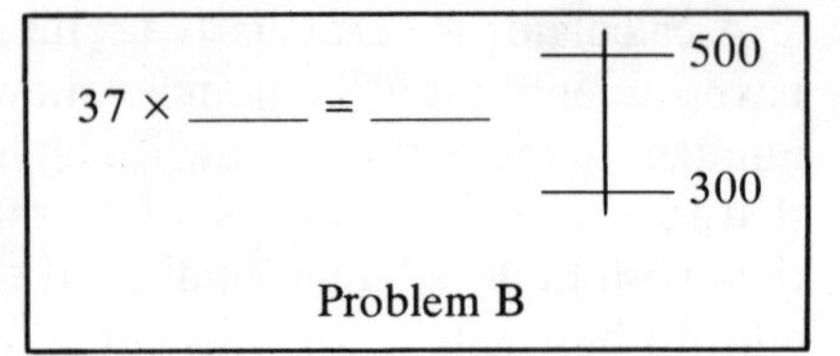

Fig. 19.3

For problem B, the solution process might be as follows:

Try 10; 37 × 10 = 370; *in the range*
Try 20; 37 × 20 = 740; *too large*
Try 15; 37 × 15 = 555; *too large*
Try 14; 37 × 14 = 518; *too large*
Try 13; 37 × 13 = 481; *in the range*
Try 9; 37 × 9 = 333; *in the range*
Try 8; 37 × 8 = 296; *too small*

Whole number solutions are 9, 10, 11, 12, and 13.

When the domain is restricted to whole numbers, problem A yields a high of 102 and a low of 52, with fifty-one possible choices. Problem B, with a much larger range, has a high of 13, a low of 9, and only five possible choices. A discussion of why this can occur will help students understand multiplication more clearly.

An interesting aspect of the activity can be seen when there are no whole numbers in the solution set (see fig. 19.4).

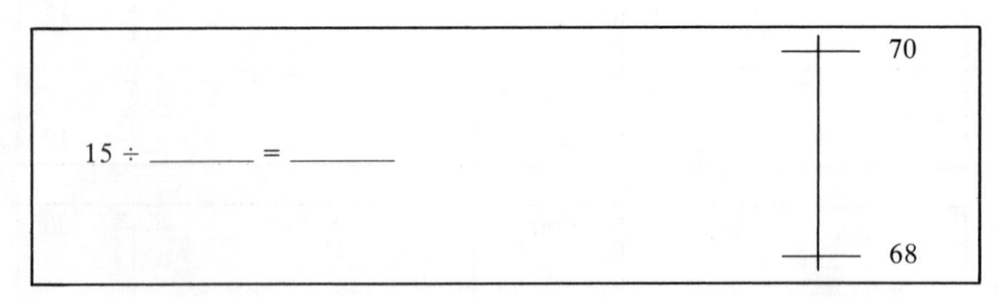

Fig. 19.4

For this problem the reasoning might be as follows:

Try 3 ; 15 ÷ 3 = 5; *too small*
Try 2 ; 15 ÷ 2 = 7.5; *too small, but closer*
Try 0.5 ; 15 ÷ 0.5 = 30; *too small*
Try 0.6 ; 15 ÷ 0.6 = 25; *oops! wrong direction*
Try 0.3 ; 15 ÷ 0.3 = 45; *better*
Try 0.2 ; 15 ÷ 0.2 = 75; *too large; answers are between 0.2 and 0.3.*
Try 0.25; 15 ÷ 0.25 = 60; *too small*
Try 0.22; 15 ÷ 0.22 = 68.18; *success!*

Students normally infer that the required number in these problems is more than 1. They often go in the wrong direction after their first feedback from trying a decimal number, but gradually they begin to gain an understanding of what occurs when one divides by numbers that are less than 1. The question, "Is there a largest (smallest) number?" proves to be quite provocative. A similar approach (fig. 19.5) can be used to introduce and explore negative integers.

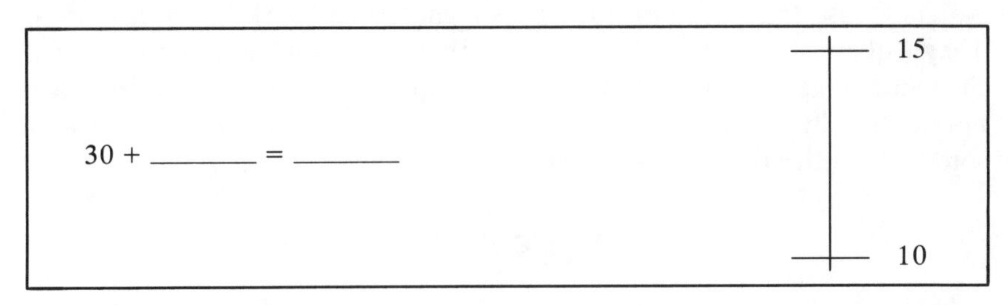

Fig. 19.5

Activities similar to this one can be found in materials by Immerzeel and Ockenga (1977) and Reys et al. (1979). The activities are easily adapted to student ability or grade level. We believe on the basis of our work in schools that estimation activities of this type should be used throughout the school year. The payoff actually increases as students become experienced with the game. With the increased awareness of the importance of estimation and with the continuing importance of concept development in mathematics, activities that enhance learning in both areas—as these do—are of obvious worth.

REFERENCES

Immerzeel, George, and Earl Ockenga. *Calculator Activities for the Classroom,* Books 1 and 2. Palo Alto, Calif.: Creative Publications, 1979.

Reys, Robert, Barbara Bestgen, Terrence Coburn, Robert Marcucci, Harold Schoen, Richard Shumway, Charlotte Wheatley, Grayson Wheatley, and Arthur White. *Keystrokes: Calculator Activities for Young Students.* Palo Alto, Calif.: Creative Publications, 1979.

20

The Computer and Approximate Numbers

Edward J. Rolenc, Jr.

STUDENTS typically believe that mathematics is always an exact world. They base this belief on their paper-and-pencil work in mathematics. They will eventually discover, however, that computers do not always give the values that they expect; that is, the computer's computational rules are not necessarily the same as those for exact numbers. This article illustrates some difficulties that may arise and presents some possible remedies.

DIFFICULTIES

Many times it is necessary to have a computer check for equality in a conditional statement. This can cause some interesting results when programming in BASIC. Here is an example of a conditional statement:

```
IF X - 1 = 0 THEN PRINT X, "Undefined"
```

The hypothesis IF X − 1 = 0 is either true or false. When the "IF" part of such a statement is true, the computer will execute the "THEN" part. When the "IF" part is false, control passes to the next line of code in the program.

The following assignment will help demonstrate to your students some problems that may arise when using equality with conditional statements in BASIC.

Exercise. Write a BASIC program for the computer that will generate a set of coordinate points for the following function over the domain $-2 \leq x \leq 2$:

$$Y = \frac{X - 4}{X - 1}$$

The following program should generate the coordinate points over the specified domain:

```
10 PRINT "Domain Value", "Range Value"
20 PRINT
30 LET X = -2
40 IF X - 1 = 0 THEN PRINT X, "Undefined"
50 IF X - 1 < > 0 THEN PRINT X, (X - 4) / (X - 1)
60 LET X = X + .1
70 IF X < = 2 then 40
80 END
```

Your students can type the program into their computers and execute it. Here are some coordinate values generated on a microcomputer (the values from your students' computers may vary slightly from those below):

Domain Value	Range Value
−2	2
−1.9	2.03448
.	.
.	.
.	.
−.2	3.5
−.0999997	3.72727
3.12924E−07	4
.1	4.33333
.	.
.	.
.	.
.9	31.0001
1	−8.38861E+06
1.1	−28.9999
.	.
.	.
.	.
1.8	−2.75
1.9	2.33333

Students will find some interesting surprises in the output. For instance, instead of printing a domain value of 0, the computer prints 3.12924E − 07. (The E signifies that the value is given in scientific notation.) Another surprise occurs when $x = 1$. Your students will expect that the conditional given in line 40 will force the computer to have the word "Undefined" in the range column, since the denominator of $(x - 4)/(x - 1)$ becomes zero when x is 1. However, the computer gives a result of −8.38861E + 06, or −8388610. Finally, notice that the domain column does not include 2, even though line 70 specifies that the computer is to continue incrementing by 0.1 as long as x is less than or equal to 2. Your students will wonder why these outcomes occur and how to remedy them.

REMEDIES

Many of these difficultues are due to approximations that result from the computer's use of the binary number system. The binary equivalent to the decimal value of 0.1, the increment in our program, is $0.000110011_2 \ldots$, a repeating binary fraction. The computer must truncate, or "cut off," the binary fractional value at some point. This introduces a rounding error into the results, and exact equality with a whole number like 1 or 2 cannot be attained.

There are several ways that programmers can circumvent these difficulties. One method is to build a margin of error (or a tolerance factor) into the program. This can be done in our earlier sample program by replacing lines 40, 50, and 70 as follows:

```
40 IF ABS ( (X – 1) – 0) < = .001 THEN PRINT X,
   "Undefined" : GOTO 60
50 PRINT X, (X – 4) / (X – 1)
70 IF X < = 2.1 THEN 40
```

The conditional statement in line 40 specifies that if the absolute value of the difference between $x - 1$ and 0 is less than or equal to 0.001, then x, "Undefined," is printed. This relaxes the strict equality of $x - 1$ and 0 in the original line 40. The unconditional GOTO at the end of line 40 sends control to line 60. If the conditional in line 40 fails, control will automatically pass to line 50.

Have your students execute this revised program. The domain and range values in the output will be identical to those obtained before the revision with two exceptions. First, the domain value of 1 will give the range value "Undefined." The margin of error built into the new line 40 has given us this desired outcome. A graph illustrating what is happening in line 40 may help your students grasp what a margin of error actually does:

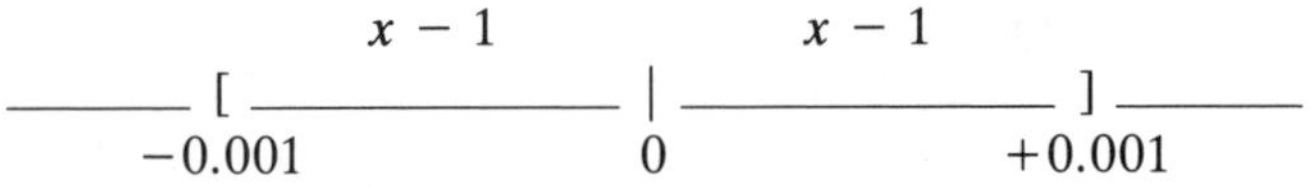

```
              x − 1                 x − 1
_______ [ ____________ | ____________ ] _______
     −0.001            0           +0.001
```

Line 40 causes the computer to print "Undefined" whenever x is between -0.001 and 0.001. This is really allowing the computer a tolerance factor, or a margin of error of ± 0.001. Some of your students may decide that they need a more restrictive margin, say, 0.0001, or a less restrictive one, say, 0.01. The appropriate margin of error will depend on the program's purpose. Students will find from experimenting that some margins are too "tight" and some are too "loose" for different purposes. The final outputs should tell them whether they have selected appropriate margins of error.

The output from our revised program also gives us a domain value of 2 and a corresponding range value of -2. By changing the X< = 2 in line 70 to X< = 2.1, the programmer causes the computer to exhibit a range value for the domain value of 2. This demonstrates the use of "padding" values in programs, another type of tolerance factor.

Many of the difficulties above could also be bypassed by using an increment value that, unlike 0.1, is an integral power of 2. One such number is one-eighth (0.125 is decimal form). Since one-eighth is represented in binary as the exact number 0.001_2, your students will find that no tolerance factors are needed when line 60 in our original program is replaced by this line:

```
60 LET X = X + .125
```

Here are some of the output values generated by this revised program:

Domain Value	Range Value
−2	2
−1.875	2.04348
.	.
.	.
.	.
.125	3.66667
0	4
.125	4.42857
.	.
.	.
.	.
.875	25
1	−Undefined
1.125	−23
.	.
.	.
.	.
1.875	−2.42857
2	−2

Notice how much "cleaner" the values seem to be. If your students feel that a value of 0.125 is not a suitable increment, then have them try other decimal values that are integral powers of 2, such as one-fourth or one-sixteenth.

SUMMARY

When your students encounter the difficulties set forth in this article, they will wonder what is wrong with their logic. In fact, the difficulties arise because of the computer's use of approximate real numbers. The world of mathematics as seen through a computer is not the exact world that most students believe it to be. This is an important lesson for students to learn, and the problem solving required to circumvent the difficulties is an added educational benefit.

21

A Systematic Guessing Approach to the Crossed Ladders Problem

Dmitri Thoro

LET us consider the infamous crossed ladders problem:

Crossed ladders of lengths a and b intersect c units above an alley (fig. 21.1). Find w, the width of the alley.

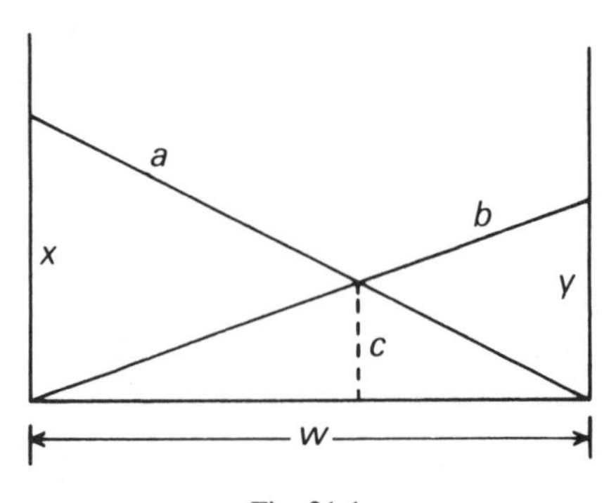

Fig. 21.1

The classical approach involves a hard-to-find polynomial equation of degree 4 (whose real roots are usually irrational). This equation can be solved by guessing an initial approximation and then employing an iterative technique (such as the bisection method, Newton-Raphson method, fixed-point iteration, etc.) to generate a sequence converging to w.

Here, however, we shall use only a little geometry, systematic *repeated* guessing, and good old-fashioned linear interpolation. This procedure can be applied to a surprisingly large number of otherwise intractable problems.

First let us note that *if* (instead) w, a, and b were given, then c would be determined. It is not hard to show that from similar triangles $1/c = 1/x +$

$1/y$, or $c = xy/(x + y)$, where $x = (a^2 - w^2)^{1/2}$ and $y = (b^2 - w^2)^{1/2}$. Thus $c = f(w)$.

Now let us take $c = 10$ as an example. Our objective will be to guess two values of w: one that yields $c < 10$ and one that yields $c > 10$. (Of course $0 < w < \min(a,b)$.) Then linear interpolation can be used. (If these values are w_1 and w_2, the interested reader can use a little algebra to show that we can interpolate linearly between w_1 and w_2 by computing

$$\begin{vmatrix} w_2 & c - f_2 \\ w_1 & c - f_1 \end{vmatrix} /(f_2 - f_1),$$

where $f_i \equiv f(w_i)$.) The entire procedure can be repeated as desired (until we obtain $f(w)$ sufficiently close to the original value of c).

Assume the original parameters are $a = 30$, $b = 20$, $c = 10$. We may proceed in the following manner:

Step 1. Start with a guess w that is close to 20 (and hence $c < 10$).

Step 2. Repeatedly *decrease* each subsequent guess until the corresponding $c > 10$.

Step 3. Then *increase* each guess (using smaller increments) until once again $c < 10$.

Step 4. Now compute the interpolated value for w.

A BASIC program based on this algorithm is shown in figure 21.2.

```
 10 REM  YOU ARE INVITED TO SUPPLY COMMENTS AND
         MODIFICATIONS
 20 INPUT A1
 30 LET W = A1
 40 GOSUB 180
 50 LET C1 = C
 60 PRINT W, C
 70 IF C < 10 THEN 20
 80 INPUT A2
 90 LET W = A2
100 GOSUB 180
110 LET C2 = C
120 PRINT W, C
130 IF C > 10 THEN 80
140 LET W = (A2 * (10 - C1) - A1 * (10 - C2))/(C2 - C1)
150 GOSUB 180
160 PRINT W, C
170 STOP
180 LET X = SQR (900 - W * W)
190 LET Y = SQR (400 - W * W)
200 LET C = X * Y / (X + Y)
210 RETURN
220 END
```

Fig. 21.2

Typical input/output follows:

Step	w	c = f(w)
1	15	8.76555
2	14	9.28480
2	13	9.72938
2	12	10.11433
3	12.2	10.04158
3	12.4	9.96678
4	12.30994	10.00072

Even though our original guess ($w = 15$) was quite crude, the procedure described yields an approximation to w that is delightfully close to the exact solution (which, rounded to three decimal places, is 12.312).

The reader is invited to show that with some minor modifications this program can be used to solve the following problem.

> Our hero, James Eyespy, is at point A (see fig. 21.3). He has just completed arrangements to rescue a damsel in distress, Penelope Blissmore, who is imprisoned at point B. For strategic reasons he cannot go directly from A to B. Instead, Eyespy will run from A to P (at an average rate of 6 km/h). There, on Highway 006, a slow (but friendly) helicopter will fly him and his rescue equipment to point B (at an average rate of 60 km/h). The success of the mission depends on arriving at B in exactly one hour. Find x.

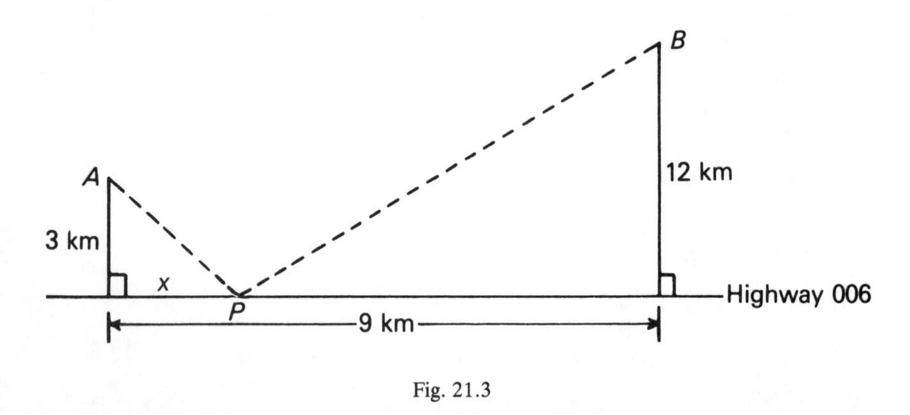

Fig. 21.3

Systematic guessing will again provide a solution, without resorting to difficult equations.

Benchmark: Estimating Bounds for Definite Integrals in Geometric Problems

Richard C. Pappas

HOW CAN students estimate results for geometrical problems whose exact solution requires the evaluation of certain definite integrals? Here is a method with three examples, each of which uses some elementary geometry to establish both an upper and a lower bound for the quantity required in the problem. Such estimation can be used to check the reasonableness of the calculated value of the integral while strengthening in the students' minds the link between the definite integral and its geometric interpretation.

Example 1. Find the length of the curve $y = x^{3/2}$ between the points $(0,0)$ and $(2,2\sqrt{2})$.

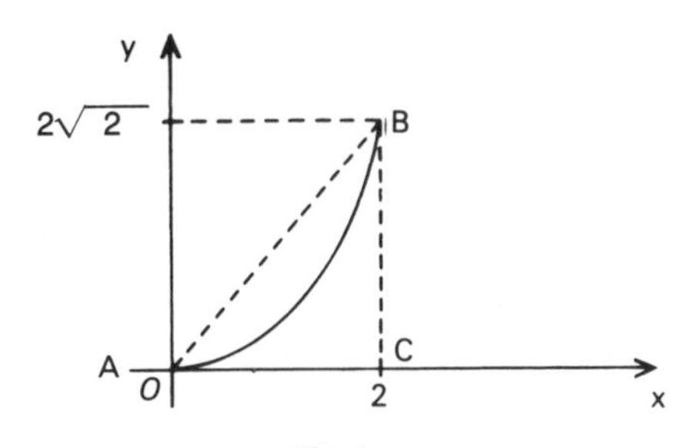

Fig. 1

The relevant part of the curve is shown in figure 1. Let s be the required length. Clearly s is longer than the hypotenuse of the right triangle ABC; but s is also shorter than the sum of the two sides $AC + CB$. Thus, s must satisfy

$$2\sqrt{3} < s < 2 + 2\sqrt{2},$$

or, to three significant figures,

$$3.46 < s < 4.83.$$

From the sketch, it also seems that s should be closer to the lower limit than the upper limit. An evaluation of the integral gives $s = 3.53$.

Example 2. Find the area under the curve $y = \sin x$ from $x = 0$ to $x = \pi$.

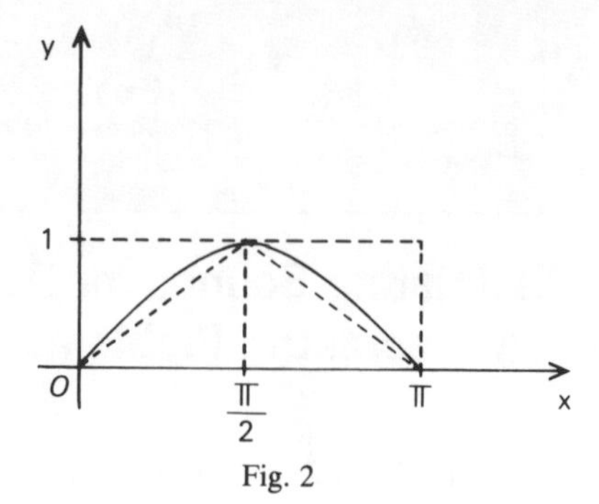

Fig. 2

If we call the required area A (fig. 2), then clearly A is less than the area of the rectangle circumscribed around the arc of the curve and greater than the area of the inscribed triangle. Thus we have $\pi/2 < A < \pi$, or, to three significant figures, $1.57 < A < 3.14$. The exact answer is $A = 2$.

Example 3. Find the volume cut from the elliptic paraboloid $z = x^2 + 4y^2$ by the plane $z = 1$.

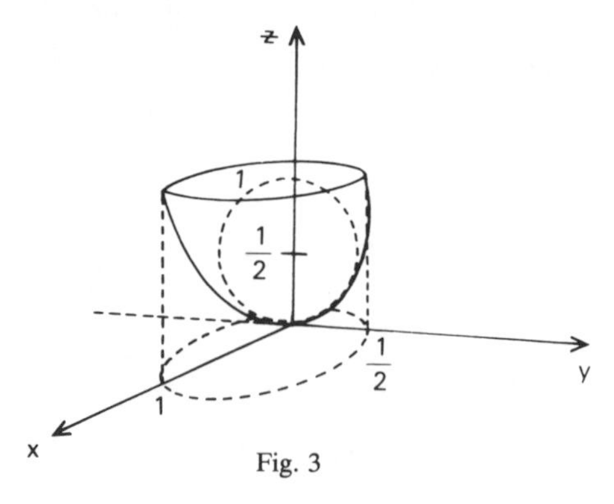

Fig. 3

As an upper bound to the required volume V shown in figure 3, consider the elliptic cylinder whose base in the x-y plane is

$$x^2 + 4y^2 = 1$$

and whose height is 1. Its volume is $\pi/2$. As a lower bound, consider a sphere of radius 1/2 with its center at $z = 1/2$. Its volume is $\pi/6$. Hence we estimate

$$\pi/6 < V < \pi/2.$$

Exact calculation of the appropriate triple integral gives $V = \pi/4$.

Although these examples are quite simple, experience has shown that almost all problems of this type that are given in freshman calculus texts are susceptible to such analysis. At times some ingenuity is required, particularly to arrive at the lower bounds. But even an upper bound alone is better than no estimate at all!

22

Estimation in Measurement

Terrence G. Coburn
Albert P. Shulte

ESTIMATION is deservedly receiving a great deal of emphasis in current curriculum planning. Techniques of computational estimation, such as front-end estimation and rounding off, are now being included in district mathematics objectives; they are being taught, and students are being evaluated on their ability to use them.

Estimation in measurement activities is equally important. Students need to develop the skill to get a ball-park figure and then a more refined estimate in measurement as much as in computation. However, measurement is usually taught by having students perform only the mechanics of measuring without the practice of estimating measurements.

In the NCTM's 1976 Yearbook, *Measurement in School Mathematics,* Bright (1976, 93) states two purposes for teaching estimation in measuring: "first, to help students develop a mental frame of reference for the sizes of units of measure relative to each other . . . and, second, to provide students with activities that will concretely illustrate basic properties of measurement." Certainly one could add a third: giving students a means for determining whether a given measurement is reasonable.

Bright lists eight basic types of estimation involving measurement, classifying them as shown in figure 22.1. He advocates that students gain experience with each of these eight situations.

There are several strategies for teaching estimation in measurement:

- Compare the object to be measured with a referent.
- Estimate, measure, and check. Have students do this enough times to improve their estimation ability, but don't let them overdo it. Be sure to point out that in many situations an actual measurement is never done—either because it's impossible or because an estimate is good enough.
- Have students attempt to estimate within a given amount or a given percent of the actual measurement. In *The Agenda in Action,* Underhill (1983, 99) recommends beginning with this *range* estimation.

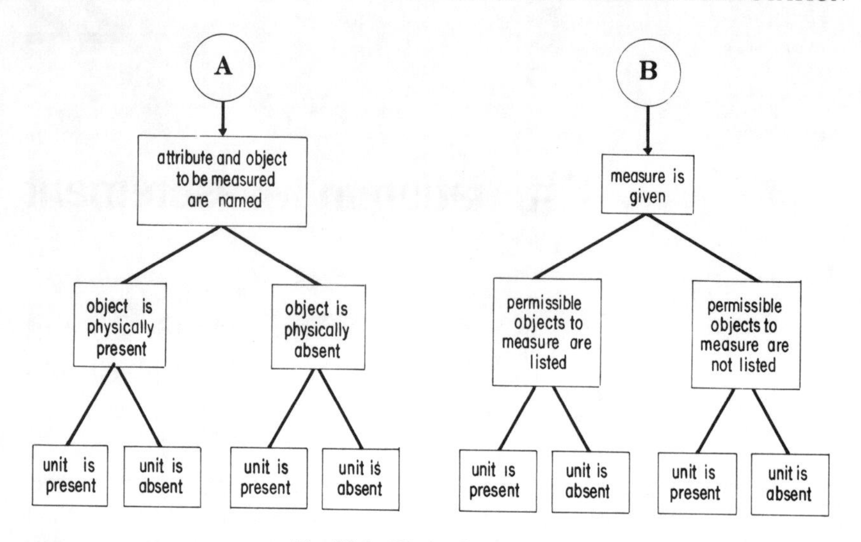

Fig. 22.1. Kinds of estimation

- Present activities and allow only a brief time for solution, thus compelling students to estimate rather than compute.

Careful planning is required, but teachers who incorporate estimation in their measurement instruction will be amply rewarded. Instruction will become more active, meaningful, motivating, and fun!

On the following pages you will find examples of activities that can be used for teaching estimation in measurement. Several are set forth in activity-card format—a format that can be particularly useful when you want students to work individually or in small groups. The others could easily be recast into the activity-card format.

Activities have been selected from the following types of measurement: (1) length, (2) area, (3) volume, (4) capacity, (5) mass, (6) time, and (7) analogy. Representative activities are included for the lower elementary, upper elementary, and middle school/high school levels.

Length **Upper Elementary**

Twenty-one

Estimate the length of the line segment on each card. Accumulate enough cards to have a total length less than or equal to 21 cm.

Materials: A deck of 20 cards (about 8 cm × 12 cm). Letter the cards on one side and draw line segments of random length on the other (one on each card).

A list of the line lengths for each card

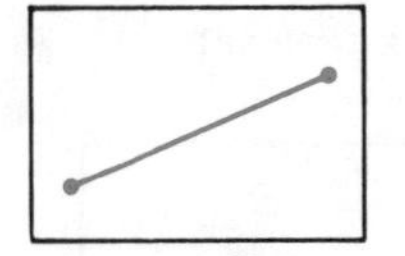
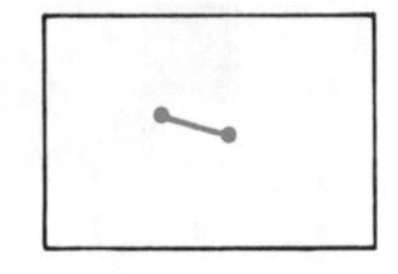
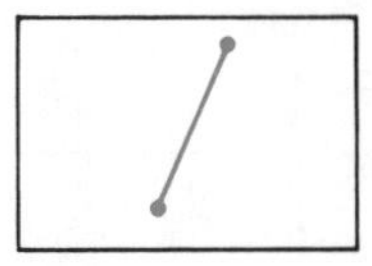

Two or more can play at a time. Dealer shuffles the cards and places them in a pile face down. Players take turns drawing one card at a time from the top until one of them estimates that he or she has accumulated line segments with a total length of 21 cm and stops the game by saying "Twenty-one!" This player may discard any one card that might push the total over 21 cm. He or she then states an estimate for the total length of the remaining segments.

The **score** is computed as follows: The difference between the estimate and the actual total plus the difference between the actual and 21. At the end of a given number of games, the scores are added and the lowest score wins.

(*Note:* Adapted from the game "Metric 21")

Variation

Use cards with various polygons on them. A student sorts the pile of cards into three stacks according to whether the perimeter is (1) much greater than a given length; (2) about equal to the given length; and (3) much less than the given length.

Length **Middle School/High School**

Cutting String

Materials

Scissors
String (about 7 or 8 meters for each student)
Meterstick or tape measure
Masking tape
Pencil (to label each piece of string)

Without standing and without using a measuring device, cut a piece of string that is as long as—

1. you are tall;
2. one meter;
3. the width of the teacher's desk.

Label each piece of string with masking tape.

Compare each length of string with the actual object.

Area **Lower/Upper Elementary**

Polygons

Estimate the area of each polygon in square centimeters.

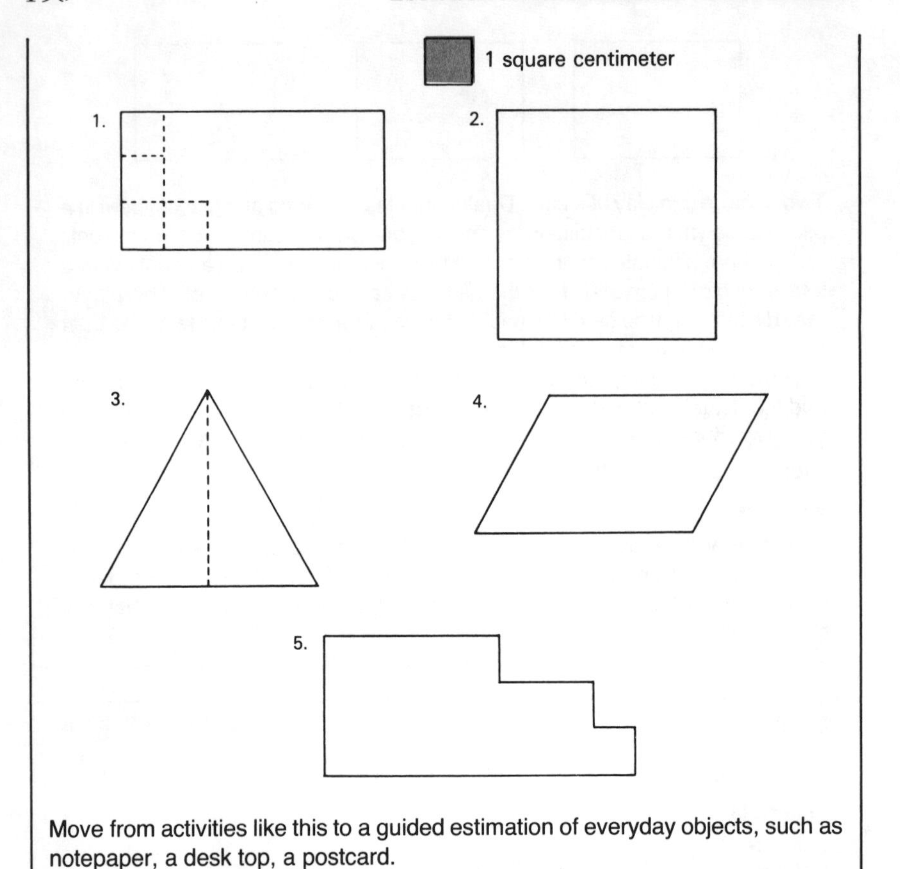

Move from activities like this to a guided estimation of everyday objects, such as notepaper, a desk top, a postcard.

Area **Middle School/High School**

Feet and Hands

1. Trace the outline of your foot on centimeter graph paper.
2. Estimate the area of your foot in square centimeters. (Either a single number or a range is acceptable.)

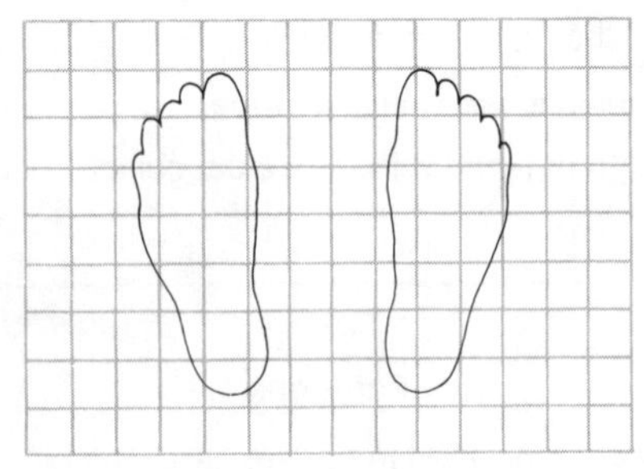

3. The area of the bottom of one's foot is roughly 1% of the body's surface area. Using the result in (2), estimate your body's surface area.
4. Trace the outline of your hand on centimeter graph paper.
5. Estimate the area of your hand.
6. About what percent of your body's surface area is the area of your hand?

Volume **Lower Elementary**

Volume with Blocks

Materials

3 or 4 open boxes (such as boxes for shoes, tape cassette, chalk, crayons, cereal)

A pile of centimeter cubes—only enough to cover the bottom of each box separately plus enough to make one stack the height of each box

A list of the actual volumes (in cubic centimeters) to use as a check

Use this pile of centimeter cubes. Estimate how many cubes it would take to fill each box. Write your estimate on the slip of paper.

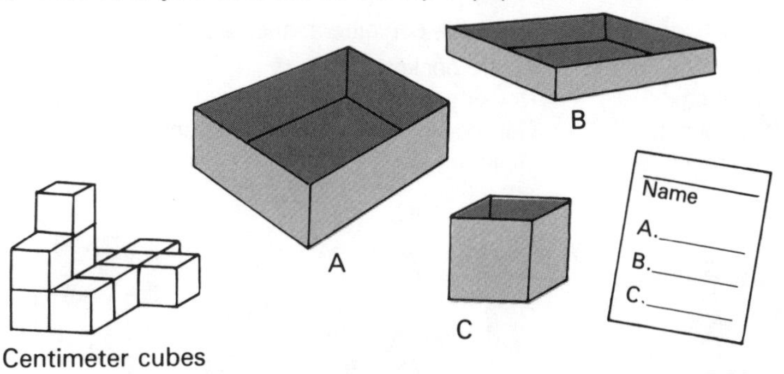

Volume **Upper Elementary**

Ordering Volume

Materials

Models of a cubic centimeter and a cubic decimeter (liter)
5 or 6 containers such as the following:

Tennis ball can	Waste basket
Shoe box	16-mm film container
Grocery bag	Plastic drinking cup
Cereal boxes	Milk carton

Each item can be lettered for identification.

Ask the students to order the containers by volume and to write their estimates for the volume of each container in both cubic centimeters and liters.

Post a list of actual volumes (the teacher's best estimates) as a check for the students' answers.

Volume **Middle School/High School**

Volume with Formulas

Estimate the dimensions and use these formulas to estimate the volume of each container.

Materials

1. Formulas for volume:
   - $V = Bh$, or $V = lwh$ (rectangular prism)
   - $V = \pi r^2 h$ (cylinder)
   - $V = 4/3\ \pi r^3$ (sphere)
   - $V = 1/3\ Bh$ (pyramid or cone)
2. Pencils, paper, string, and scissors. Do *not* permit rulers or any visual access to standard units, such as millimeter, centimeter, meter, inch, or foot.
3. A variety of commonly available containers, such as—

Tennis ball can	Waste basket
Shoe box	Drinking cup
Cereal boxes	Cardboard box for duplicating paper
Grocery bag	Desk drawer

4. A list showing the actual volume of each container (or the teacher's best estimate) to use as a check of the students' answers.

Encourage students to round the numbers used in their computations.

Capacity **Lower Elementary**

How Much Will It Hold?

Estimate the number of liters each of the objects listed would hold. Try to come within the range allowed.

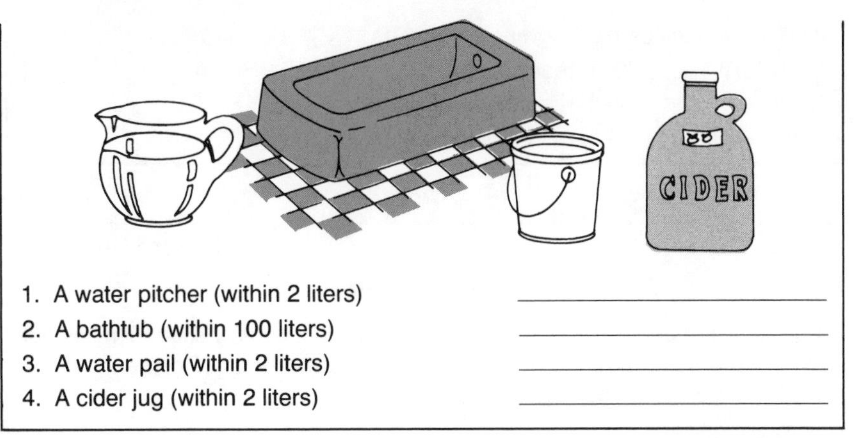

1. A water pitcher (within 2 liters) ______
2. A bathtub (within 100 liters) ______
3. A water pail (within 2 liters) ______
4. A cider jug (within 2 liters) ______

Acceptable range for water pitcher, 1–4 liters; bathtub, 400–600 liters; water pail, 6–10 liters; cider jug, 2–6 liters.

Capacity **Upper Elementary**

Finding 50 Liters

Which of the following objects has a capacity closest to 50 liters?

1. An oil drum?
2. An auto gas tank?
3. A watering can?
4. A gasoline truck?

Capacity **Middle School/High School**

Guess and Pour

This is designed as a laboratory activity. Provide such containers as a milkshake cup, a large drinking glass, a root beer mug, a soft drink bottle, a coffee cup. Number them.

1. Arrange containers 1–5 in line, in increasing order of capacity. *Do this by using your judgment, without measuring.* Record the order.
2. Check your estimates by comparison (pouring from one container to another) or actual measurement. Rearrange the containers as necessary.

3. Rate yourself on the following scale:

SUPER!	(0–1 rearrangements)
OK	(2–3 rearrangements)
Not so hot!	(4–5 rearrangements)

Mass **Lower Elementary**

Hold and Heft

This is a laboratory activity. Provide five objects of markedly different mass but roughly the same size—for example, a shoe, a large tin of canned food, a chalkboard eraser, a box of chalk, a roller skate.

1. Take 2 objects.
   Put one in each hand.
   Decide which is heavier.
2. Continue to compare objects in pairs until you have placed all 5 objects in order, lightest to heaviest.
3. Check the answer card to see how well you did.

The teacher can prepare an answer card that lists the given objects in order from lightest to heaviest.

Mass **Upper Elementary**

Matching Masses

Match the objects with their approximate mass.

a truckload of coal	3 g
a bag of sugar	30 g
a spoonful of salt	300 g
A penny	3 kg
A fifth-grade student	30 kg
A textbook	300 kg
A tiger	3000 kg

Mass **Middle School/High School**

Guess Your Weight?

Set up a "weighing station" at the school fair. Offer a prize if you can't guess a person's weight within 5 kg.

Time **Upper Elementary**

Coffee Can Clock

Punch a small hole near the bottom of an empty two-pound coffee can and place a plastic ruler inside the can. Fill with water and allow the water to empty slowly into the sink or a large pail. Notice on the ruler how much the water level drops at each (approximately) 5-second interval. Dry the can and mark the inside to correspond. Now the lowering water level should be a good indicator of elapsed time.

Variation

Burn a birthday candle and measure its height with each passing of a fixed interval of time (say, 1 minute). Make a cardboard strip that represents the length of the candle at 1-minute intervals. Burn a similar candle and check the height against the cardboard strip.

Analogies: Time/Distance **Middle School/High School**

Timing Walks to Get Distance

A person walking at "average" speed can walk a kilometer in about 10 minutes. Use this information to estimate the following distances:

1. The length of the school hall.
2. The length of a city block.
3. The distance around your yard, or the lot on which your home is located.
4. The length of the basketball court.
5. The distance from your home to your school (if you walk).
6. The distance from your house to your friend's house.

Check your estimates if possible.

REFERENCES

Bright, George W. "Estimation as Part of Learning to Measure." In *Measurement in School Mathematics,* 1976 Yearbook of the National Council of Teachers of Mathematics, pp. 87–104. Reston, Va.: The Council, 1976.

National Council of Supervisors of Mathematics. "Position Statements on Basic Skills." *Mathematics Teacher* 71 (February 1978): 147–52.

Underhill, Bob. "Estimation and Reasonableness of Results." In *The Agenda in Action,* 1983 Yearbook of the National Council of Teachers of Mathematics, pp. 97–104. Reston, Va.: The Council, 1983.

23

Teaching Measurement Estimation through Simulations on the Microcomputer

Vivian Rose La Ferla Morgan

IF YOU have bumped the car behind you while parallel parking, then you need to develop your estimation skills! Most judgments about length and area are made without a measuring instrument. Measurement estimation skills are important because (1) many practical situations do not permit the use of measurement tools and (2) estimates provide a check on the results of exact measurement. Furthermore, in real-world settings, an estimate is usually easier to obtain than a precise measurement and just as useful. Including measurement estimation in the curriculum was recommended as early as 1893 by the Committee of Ten:

> The child should learn to estimate by the eye and to measure with some degree of accuracy the length of lines, the magnitude of angles, and the areas of simple figures. (Trafton and Le Blanc 1973, p. 16)

More recently, the need for more estimation instruction has been addressed by the National Council of Teachers of Mathematics (1980) in its *Agenda for Action* and by the National Council of Supervisors of Mathematics (1977).

In addition to improving estimation skills, instruction in estimation is likely to enhance the learning of measurement concepts. According to Bright (1976), estimation activities help students develop a mental frame of reference for the size of units of measure relative to each other and to objects in the real world; and they provide illustrations of the basic properties of measurement.

Bright suggested that students should estimate the measurement of objects before measuring them and that they should keep a record of their estimates. In 1976 this meant paper-and-pencil records, but today student estimates can be stored and retrieved using a microcomputer. A microcomputer can also be used to simulate a real-world environment where the student estimates the measurement of an object. The computer compares the estimate to the actual measurement stored in the computer's memory. A student can continue to refine estimates of a measurement until a reasonable one is obtained.

The microcomputer can deal with a range of answers quickly and provide immediate feedback to a student response. Its visual display capability gives a student the means of comparing an estimate with the actual measurement. A microcomputer overcomes one of the major drawbacks of estimation instruction—the excessive time required of the teacher to evaluate student estimates individually and give appropriate instruction to improve them. The three microcomputer simulations described below can make measurement estimation both enjoyable and feasible for classroom use.

ESTIMATION SOFTWARE

These microcomputer estimation programs use an interactive, guess-and-check approach to estimation instruction. Their main objective is to facilitate the learning of measurement estimation. The programs provide instruction and practice in estimating (1) length ("Kitt"), (2) length with a scale ("Vacation"), and (3) area ("Floor Plan"). The programs are sequenced in order of difficulty but could be used separately depending on the student's level of estimation skills.

Program 1: "Kitt"

The first program, "Kitt," provides practice in linear estimation. "Kitt" simulates a map with five cities. The student travels to various cities but must start and end the trip in city 1.

As an illustration, consider the following scene. After booting the disk and typing in his or her name, the student is greeted with the following message:

YOU ARE A ROAD INSPECTOR.

YOU HAVE BEEN TOLD BY YOUR BOSS
TO INSPECT THE ROADS BETWEEN
THE MAJOR CITIES BUT ONLY
THE ROADS THAT MAKE UP THE
SHORTEST ROUTE.

YOUR CAR COMPUTER, KITT, WILL
SELECT THE SHORTEST ROAD
BETWEEN TWO CITIES BUT
YOU HAVE TO TELL IT THE
CITY AND THE *DISTANCE*.

YOUR MISSION WILL BEGIN
IN CITY 1 AND END IN CITY 1.

PRESS <RETURN> TO CONTINUE
PRESS <ESC> TO END------->

After the student presses RETURN, the map appears on the computer screen. In order to travel, it is first necessary to select a city, as demonstrated in figure 23.1. When the student selects city 2, he or she must estimate the distance from city 1 to city 2 and enter the estimate. Suppose the first

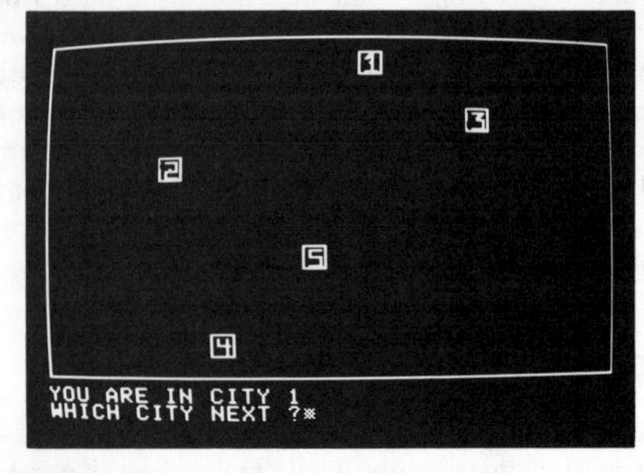

Fig. 23.1

estimate is "35." The computer draws a line on the screen representing this distance (see figure 23.2). The estimate remains on the screen for approximately two seconds, time enough for the student to examine the visual representation of the estimate but not long enough to actually measure the line segment with a ruler or straightedge. The student then uses the visual representation to arrive at a better estimate. At the bottom of the screen, a message interpreting the response is displayed:

YOUR ESTIMATE OF 35 IS TOO LOW.
ENTER VALUE AND PRESS <RETURN>.

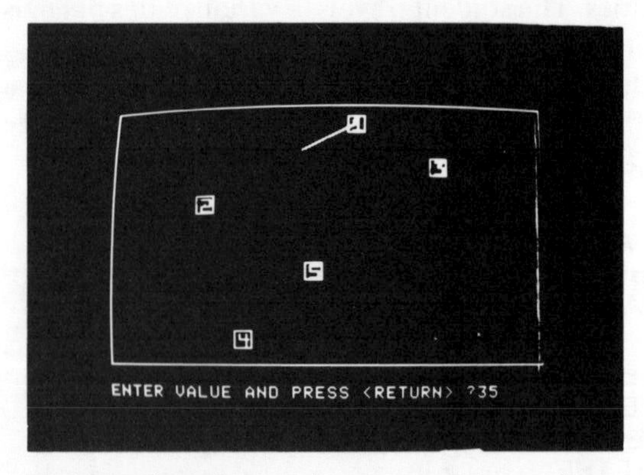

Fig. 23.2

With each estimate the computer continues to give visual and verbal feedback until an estimate is considered "close enough." The program judges an estimate to be close enough when it is within 10 percent of the exact answer. When the student responds within the correct range, the computer displays the following message:

GREAT ESTIMATE!!
TO THE NEAREST HUNDREDTH
THE MEASUREMENT IS 101.53

In response to the exact answer, the computer displays the following message:

GREAT ESTIMATE!!
101.53 IS THE CORRECT MEASUREMENT
TO THE NEAREST HUNDREDTH

Once the student has successfully estimated the first distance, the estimated length remains on the screen, since it represents a distance traveled and can be used as a measure for other estimates (figure 23.3). Now the journey can continue to the other cities until the traveler finally arrives back at city 1. This completes the simulation, and the student's estimates are recorded on disk for the teacher to examine. Each student can have an individual diskette, or a single diskette can store the student's estimates for the entire class. The student has an option to try again when the computer displays the following message:

YOUR MISSION IS OVER.
WOULD YOU LIKE TO TRY AGAIN?
TYPE <Y> OR <N>

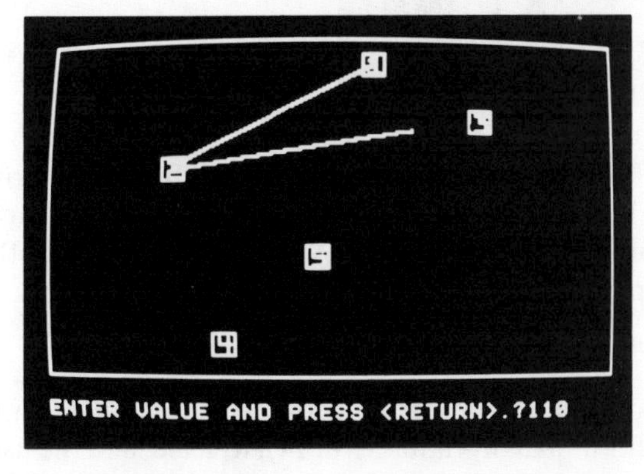

Fig. 23.3

When a student elects to try again, the scale will remain the same but the cities will be randomly placed at different locations in order to present a different problem.

Program 2: "Vacation"

Like "Kitt," "Vacation" gives practice in linear estimation, but now a scale is included (figures 23.4 and 23.5). The scale is given for the first guess and remains the same for the entire vacation trip. However, each time the student goes on a new vacation, a new scale is randomly generated. The range of possible values for the scale is 10 to 900 miles per inch. The varying

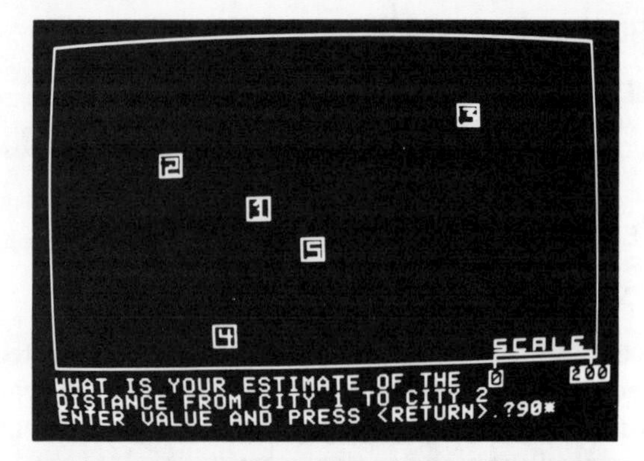

Fig. 23.4

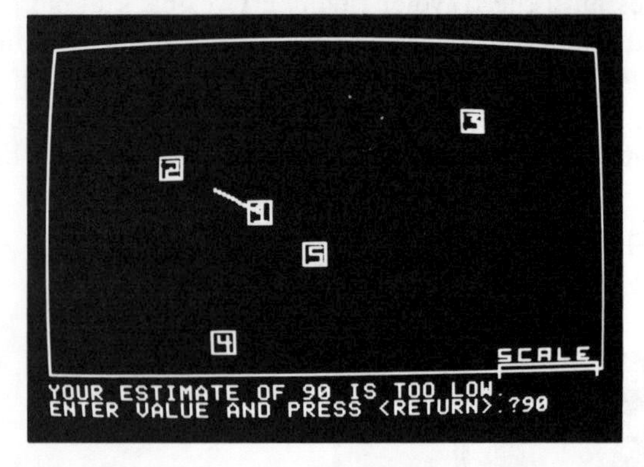

Fig. 23.5

scale requires the student to make a perceptual adjustment from one problem to the next. Thus, even though the unit of measure is given, the "Vacation" program is more challenging than "Kitt." With the exception of the scale and the theme, the procedure is the same as in "Kitt."

Program 3: "Floor Plan"

The third program, "Floor Plan," provides practice in estimating areas of simple plane figures. The floor plan of a house is drawn to scale, and the user must select both the type of flooring and the amount needed for a specific room. Consider the following scenario:

> MR. E.T. LUNA HAS HIRED
> YOU TO BE A CONTRACTOR'S
> ASSISTANT AT LUNA CITY.
>
> YOUR FIRST JOB IS TO
> SELECT THE TYPE AND
> AMOUNT OF FLOOR COVERING
> FOR MR. LUNA'S HOUSE.
>
> YOU WILL BE GIVEN A
> FLOOR PLAN OF THE HOUSE

AND CAN BEGIN YOUR
WORK IN ANY ROOM.
PRESS <RETURN> TO CONTINUE
PRESS <ESC> TO END ------>

After the student presses RETURN, a floor plan of the house appears with each room numbered (figure 23.6). The student selects a room and the type of flooring for the room. Each type of flooring (rug, tile, no-wax floor, and flagstone) represents a different unit of measure, although all are either

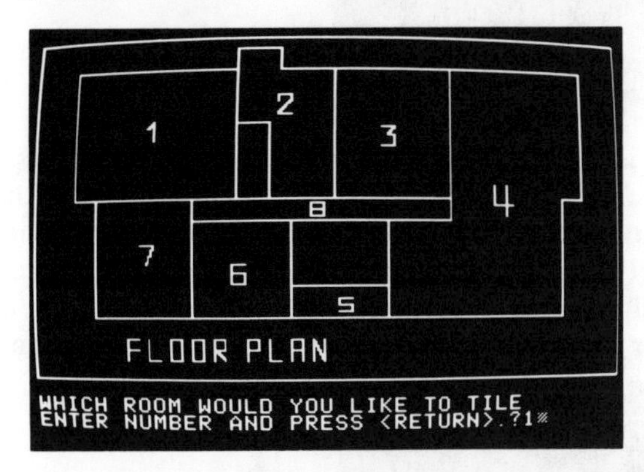

Fig. 23.6

square or rectangular. The selections are given in order of difficulty. The first unit, the rug, is a small square unit. Multiple units of rug can be placed to cover each simple figure. With a minimum number of tries, the student can obtain the exact number of units of rug needed to cover any room (figures 23.7 and 23.8). However, the third unit of measure, no-wax floor, is a larger square unit and does not cover the plane figures exactly, so the student is forced to use his or her estimation skills to arrive at a "good" estimate.

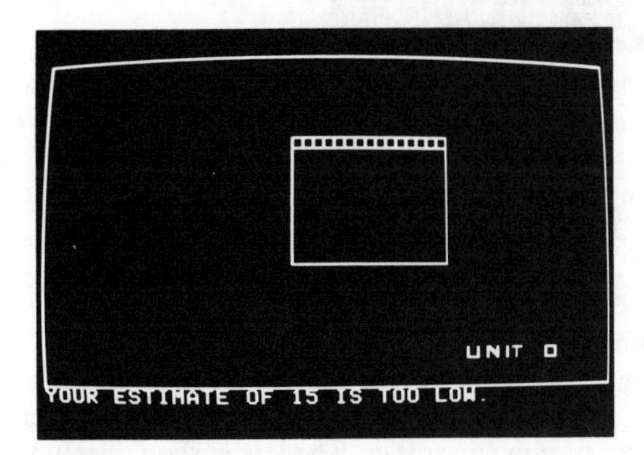

Fig. 23.7

For example, if a student selects room 1 and no-wax floor he or she must estimate the area of room 1 using a large square as the given unit of measure.

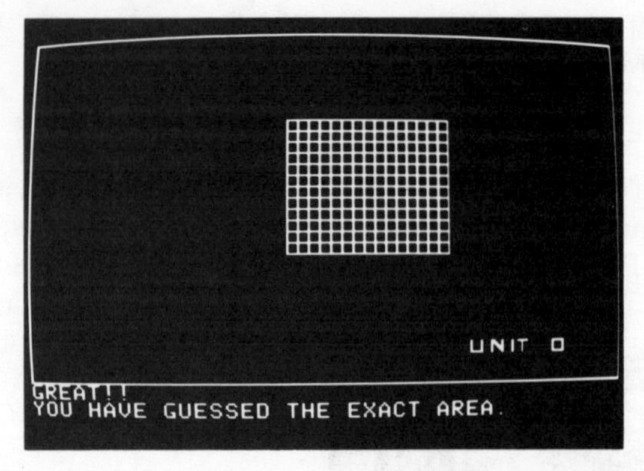

Fig. 23.8

Suppose the first estimate is "17." The computer then draws seventeen large square units on the plane figure (figure 23.9). The estimate remains on the screen for approximately two seconds, enough time for the student to revise his or her mental image of the size of the unit but not long enough to measure the figure. Using the results of previous estimates, the student continues to enter estimates until he or she is within one unit of the exact measurement.

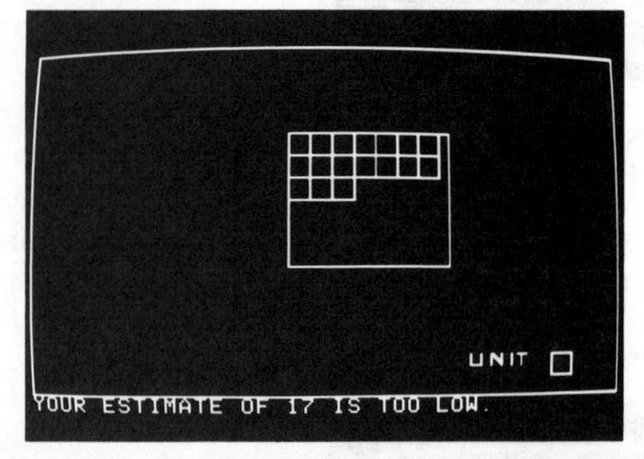

Fig. 23.9

Once the user has estimated the area of one room, he or she can select another room and a different type of flooring. For example, a student who elects to tile room 3 will be given a rectangular unit of measure. If the first estimate is "19," the computer will draw nineteen rectangular tiles on the plane figure (see figure 23.10).

IMPLEMENTING THE SOFTWARE

These programs require an Apple II Plus microcomputer with 48K of memory, monitor, and one disk drive. To operate the microcomputer and software, minimal training is necessary. An introductory lesson for students not familiar with the Apple system will suffice. The lesson should include a discussion of the hardware components, proper handling and use of disks, and the keyboard.

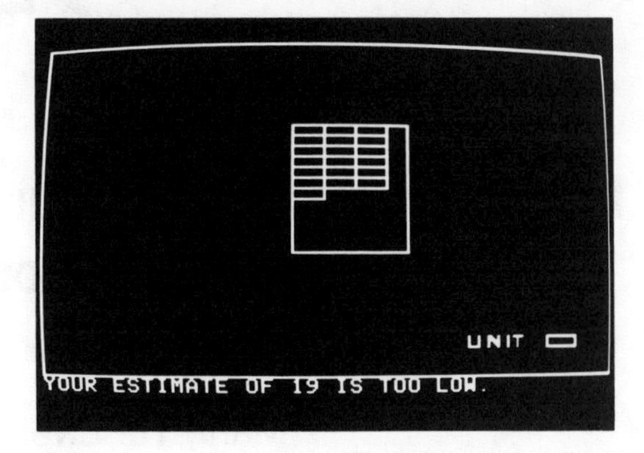

Fig. 23.10

CONCLUDING REMARKS

The software described makes it feasible for teachers to incorporate additional practice in measurement estimation into the mathematics curriculum with ease. The microcomputer is an ideal vehicle for enhancing measurement estimation instruction because (1) its dynamic graphics capability can help students develop a mental frame of reference for the sizes of units of measure; (2) it can judge whether an estimate is within a particular interval; (3) it can provide immediate feedback; and (4) it can store and retrieve student responses. Improvements in microcomputer technology, especially in the graphics area, will enhance future estimation software by allowing for a more detailed and animated screen.

Note: Information concerning the measurement estimation software described here can be obtained by writing to the author at Rhode Island College, Mathematics and Computer Science Department, Providence, RI 02908.

REFERENCES

Bright, George W. "Estimation as Part of Learning to Measure." In *Measurement in School Mathematics,* 1976 Yearbook of the National Council of Teachers of Mathematics, pp. 87–104. Reston, Va.: The Council, 1976.

Trafton, Paul R., and John F. Le Blanc. "Informal Geometry in Grades K–6." In *Geometry in the Mathematics Curriculum,* Thirty-sixth Yearbook of the National Council of Teachers of Mathematics, p. 16. Reston, Va.: The Council, 1973.

National Council of Supervisors of Mathematics. "Position Paper on Basic Skills." *Arithmetic Teacher* 25 (October 1977): 19–22.

National Council of Teachers of Mathematics. *An Agenda for Action: Recommendations for School Mathematics of the 1980s.* Reston, Va.: The Council, 1980.

24

Introducing Angle Measure through Estimation

Marilyn J. Zweng

MOST middle school teachers have experienced the frustration of teaching angle measure. They have watched as protractors are used upside down and backward. They have seen the protractor's circumference laid on top of a vertex. They have heard angles identified with measures that have no relationship to the angles in students' exercises. They spend almost all their time teaching gimmicks for reading protractor scales rather than teaching mathematics.

It appears that angle measure is difficult for students for at least three reasons. First, many students have no "sense" of angle size. They aren't the least bit surprised when they measure an obtuse angle and get 38° or 65°, or an acute angle and get 135° or 172°. "Sense of size" implies, for angles, an ability to estimate—just as it does for linear measure and area. For angle measure, however, estimation is an even more critical skill than for other measures. The device for measuring angles, the protractor, is a difficult instrument to use; consequently, for most students, estimates serve as a *check on the proper use of the measuring tool.* This is unlike, for example, linear measure in which the measuring tool typically serves as a *check on the estimate.*

Second, the protractor is, to most youngsters, a meaningless instrument. After all, why is a semicircle used to measure angles? The device used to measure line segments, a ruler, is itself a line segment. This is logical; but no such analogy is apparent between angles and the angle-measuring device.

A third, less obvious reason for the problems students encounter with angle measure is the lack of an appropriate definition for *degree.* The definitions in textbooks may be mathematically appropriate (although this is not always true), but they are certainly not satisfactory from a pedagogical perspective. For example, a contemporary seventh-grade book states: "The unit measure for an angle is a degree." This "definition" does not give youngsters much of a clue about the nature of a degree, nor does the following definition from a recent sixth-grade text: "Angles are measured using a small angle called a degree." It then goes on to say that the number of unit angles that "fit" a given angle is the angle's measure.

Using a unit angle to measure angles is analagous to using a unit line segment to measure line segments and a unit region to measure regions. The unit angle concept, even though it seems to give a unified approach to measurement instruction, is not very satisfying for middle school students. For one thing, an angle is not finite like a line segment or a region. In fact, we work very hard trying to convince students that the rays forming the sides of an angle extend without end. Further, when measuring an angle one is not measuring the object itself, that is, the two rays that form the angle, but rather "empty space" between the rays. Finally, although the idea of a "unit angle" is sometimes thought to be a concrete entity for children, it is not. It is almost impossible to draw a unit angle so it can actually be visualized, and the angle is obviously too small a unit to be physically used as a measuring tool. Estimating angle measure is certainly not facilitated by the unit-angle approach, nor does it clear up the enigma of why protractors are shaped like semicircles.

NEEDED MODIFICATIONS

Although the protractor appears to be an illogical device to most students, the reason for its semicircular shape is obvious to the more mathematically sophisticated. When the protractor is placed correctly, the angle being measured is a central angle of the protractor's circle. The measure of the arc intercepted by a central angle is, by definition, the angle's measure—and vice versa. When using a protractor, we *read directly* the intercepted *arc*'s measure, and by virtue of this definition, *infer* the *angle*'s measure. See figure 24.1. Since a protractor measures angles by arcs, it makes a great deal of sense to teach arc measure *before* teaching angle measure—a reversal of the conventional sequence.

If angle measure is to be defined in terms of arc measure, a suitable definition of *degree* in terms of arc measure is needed. By convention, a

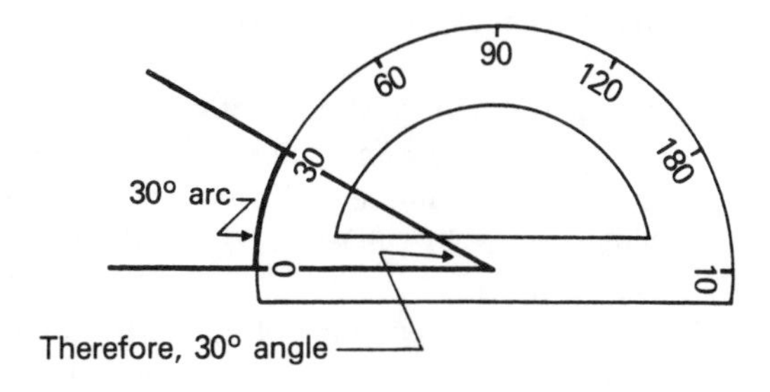

Fig. 24.1. Measuring an angle by its intercepted arc

circle is divided into 360 equal arcs. The measure of one of these arcs is defined to be 1 degree; thus, 1 degree is 1/360 of a circle. According to this definition the degree symbol (°) becomes a shorthand notation for a fraction with a denominator of 360. This is much like the use of the percent symbol (%) to indicate an implied denominator of 100. Just as 45%, for example, is a shorthand notation for 45/100, 45° is a shorthand notation for 45/360.

A degree, then, is a very different kind of measurement unit than, for example, an inch or a square centimeter. A degree is a *ratio,* a fractional part. It is not something that in multiples can be laid end to end or side by side and counted as inch segments can or square centimeter tiles. This difference between the nature of the unit used for arc and angle measurement and other units of measure is a crucial concept but one that is almost never addressed in standard curriculum materials. It is particularly crucial to teaching estimation; conversely, estimation activities help to clarify students' understanding of angle measure.

A TEACHING SEQUENCE

This section describes an instructional sequence with the following features:

- Arc measure is taught before angle measure.
- A degree is defined to be 1/360 of a circle.
- Estimation of arc and angle measure is taught *before* the protractor is introduced.

Lesson 1. Arc Measure through Estimation

Lesson 1 introduces arc measure, but estimation, only, is taught. Protractors are not used. Students are taught how to estimate a given arc's fractional part of a circle. Rather than letting them simply guess, though, the teacher encourages them to divide their circles into either sixths or eighths. For sixths the conventional construction with a compass is taught. Once the circle is divided into sixths, twelfths can also be estimated. For eighths, perpendicular diameters are drawn and each of the resulting four arcs is divided into two parts "by eye." Perpendiculars are drawn by simply using the corner of a stiff card—for example, a three-by-five-inch index card. Students may use both techniques in a given exercise if their first try does not give a particularly good estimate.

In the exercise in figure 24.2, the estimate obtained by dividing the circle into sixths is "a little more than 2/6, or 1/3, of a circle." When the circle is divided into eighths, the estimate is improved. It's "about 3/8 of the circle."

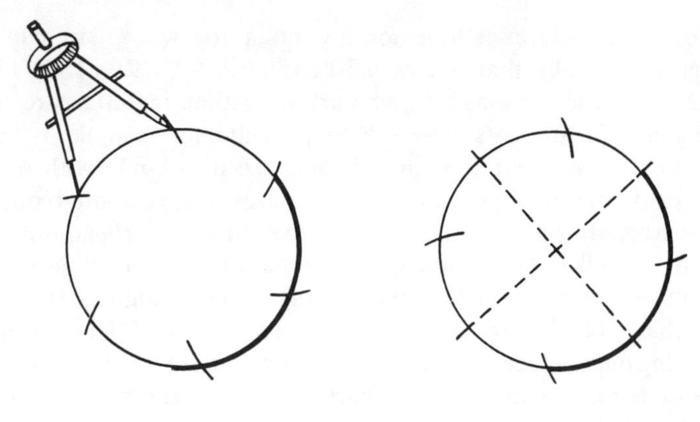

Fig. 24.2

For either construction, dividing into sixths or dividing into eighths, one of the end points of the given arc should serve as the first of the division marks on the circle.

Although the construction techniques used in lesson 1 are not needed to judge fourths and halves of circles, most students, initially, need a crutch for estimating thirds, sixths, and eighths. As a result of this first lesson, however, they quickly acquire the ability to identify arcs—without using constructions—that are approximately 1/12, 1/8, 1/6, and 1/3 of a circle and their multiples. Arcs greater than one-half of a circle are also included in the exercises, but this is not an essential feature of the lesson.

Lesson 2. Arc Measure in Degrees through Estimation

Lesson 2 introduces the degree as a unit of arc measure. As in lesson 1, all measurements are estimated. A degree is defined as 1/360 of a circle. Once an arc's fractional part is estimated, equivalent fractions are used to determine its degree measure. For example, the arc in figure 24.3 is first estimated to be approximately 1/3 of a circle. Then the fraction equivalent to 1/3, with a denominator of 360, is calculated. Mental calculation is encouraged.

$$\frac{1}{3} = \frac{120 \times 1}{120 \times 3} = \frac{120}{360}$$

$$\frac{120}{360} = 120°$$

The measure of the arc is about 120°.

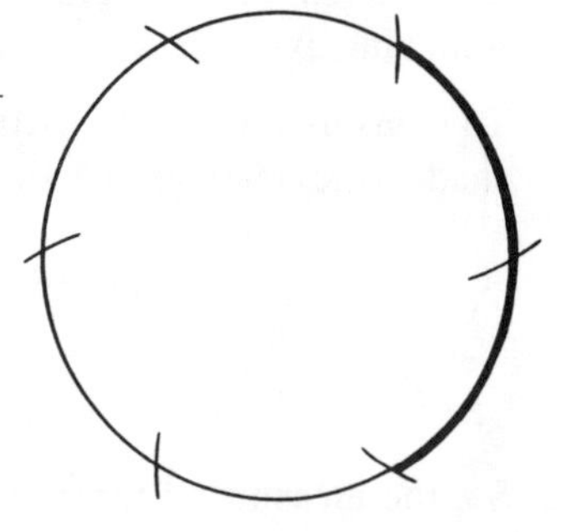

Fig. 24.3

Most of the exercises in lesson 2 contain arcs whose size can easily be recognized visually, that is, arcs of 30°, 45°, 60°, 90°, 120°, 135°, 180°, 240°, and 270°. These serve as a benchmark for estimating other arcs. The ultimate goal is for students to be able to associate, for example, 1/4 of a circle with 90°; 1/8 of a circle with 45°, and 1/6 of a circle with 60° without paper-and-pencil computation. Oral exercises are an essential component of this instruction. Pictures of arcs are shown on the overhead projector, and students are asked to estimate the fractional part and to tell the corresponding number of degrees. They are encouraged to use language such as "a little more than 1/4 of a circle" or "a little less than 60°." This lesson and the preceding one include sets of exercises in which several arcs have the same measure but different lengths. Figure 24.4 is an example. Exercises of this sort reinforce the notion that arc measure is *not* a measure of length but is a fractional part of a circle.

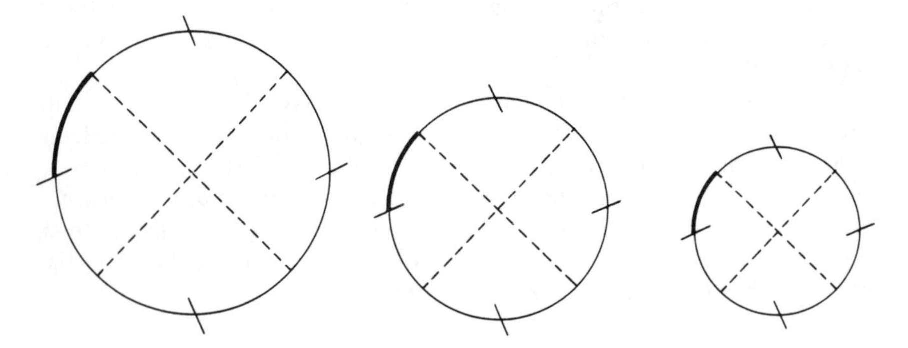

Fig. 24.4. 45° arcs with different lengths

Lesson 3. Angle Measure through Estimation

Lesson 3 introduces angle measure, but protractors are still not used. Like lessons 1 and 2, this lesson also relies on estimation. To measure an angle, students draw a circle using the vertex of the angle for its center, as in figure 24.5. Then they estimate the measure of the intercepted arc as in the following example:

To measure angle *ABC*, draw a circle using *B* for its center (fig. 24.5). The shaded arc between the sides of the angle is about 3/8 of the circle.

$$\frac{3}{8} = \frac{45 \times 3}{45 \times 8} = \frac{135}{360}$$

$$\frac{135}{360} = 135°$$

So, the measure of the angle is about 135°.

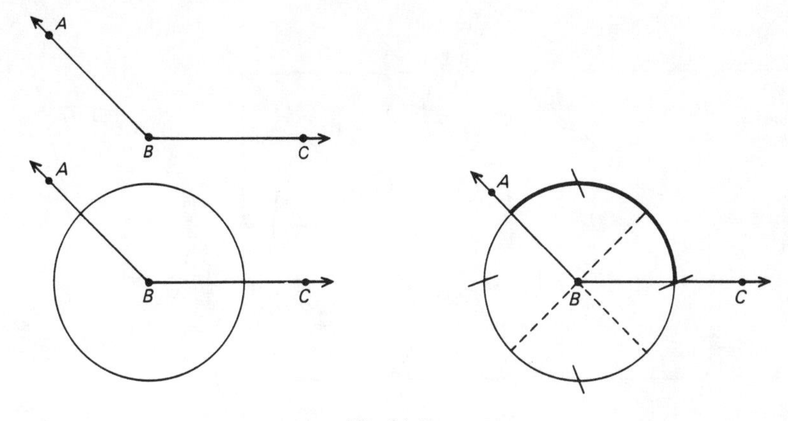

Fig. 24.5

In this lesson, constructing the circle about the vertex of the given angle is the crucial step. Once this is done, the lesson simply repeats the activities of the prior lessons on arc measure. Oral exercises are again an important follow-up to the worksheet activities. Particular attention is given to using a 90° angle as a basis for comparison by posing questions such as, "Is angle *ABC* smaller or larger than a 90° angle? Is it about half of a 90° angle? Is it halfway between 90° and 180°?" Students do become adept at estimating angle size; however, it should be pointed out that many students must continue to draw a circle about the vertex of their angles for quite some time after lesson 3. It seems as if space between two rays of an angle is difficult to deal with, whereas an arc is something more concrete and therefore easier for students to grasp.

These materials allow angles equal to or greater than 180°. Since angle measure is defined as a fractional part of a circle, the restriction to angles less than 180° is not necessary.

Lesson 4. Protractors

Up to this point, students have learned the meaning of arc measure and angle measure and a definition for degree, but they still have not been introduced to a protractor. All measurements have been estimated. Now, teaching them to use protractors is easy. If time permits, the protractor is introduced by having students construct their own. A circle is drawn, folded into eighths, and then divided into sixths with a compass. Divisions of 15° can then be obtained by measuring the distance between, for example, the 45° and 60° marks with a compass and then striking this distance off repeatedly around the circle's circumference. This gives the marks for 15°, 30°, 75°, 105°, and so forth (fig. 24.6).

Finally a "real" protractor is introduced, but this protractor is circular, not

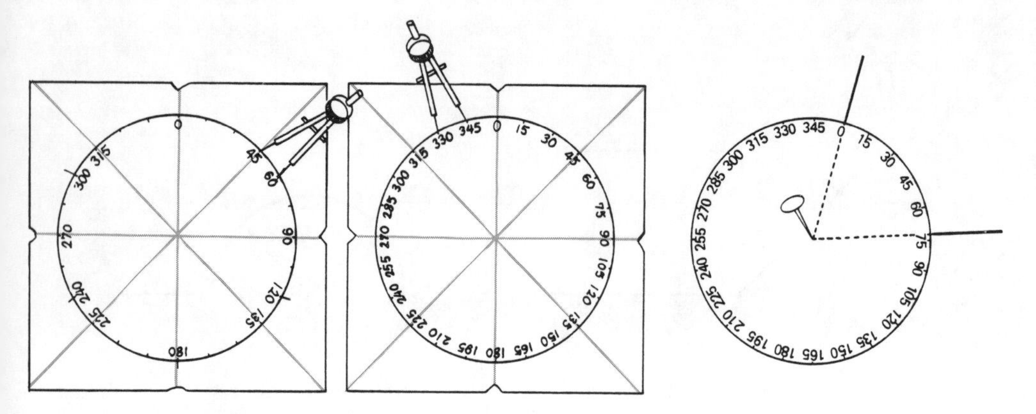

Fig. 24.6. Constructing a protractor

the more typical semicircular protractor. Initially no directions are given for using it. Students already know, however, that to measure an angle they draw a circle using the vertex for the center. The only difference with a protractor is that the circle is already "drawn" for them. Without instruction they will place the protractor so its center and the vertex of the angle coincide. Methods of reading the scale are *not* taught. Some students will place the protractor as shown in figure 24.7 and subtract 70 from 92 to find the measure of the arc, but they quickly discover that it's much easier to place the 0° mark on one of the angle's sides and read the arc measure (and hence the angle measure) directly from the protractor.

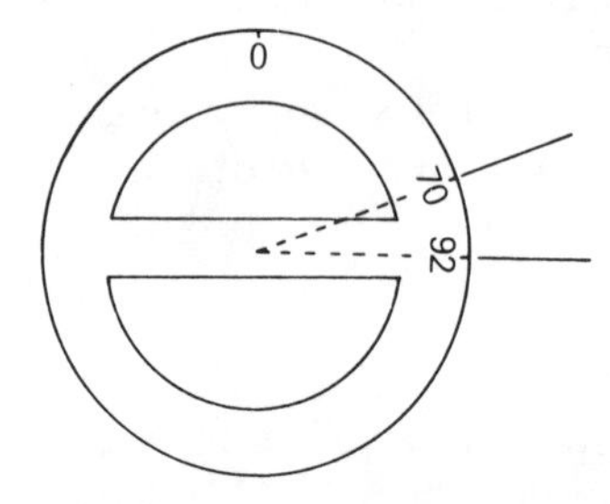

Fig. 24.7. A less efficient but correct angle-measurement procedure

One of the hazards, of course, is getting the "wrong" angle side on the 0° mark like the example in figure 24.8. However, by now, students have such a good sense of angle measure from their estimating experience that they will either reposition the protractor or calculate the difference between 325° and 360° to get the answer. They know immediately that 325° is an unreasonable answer.

The transition to the conventional semicircular protractor is almost au-

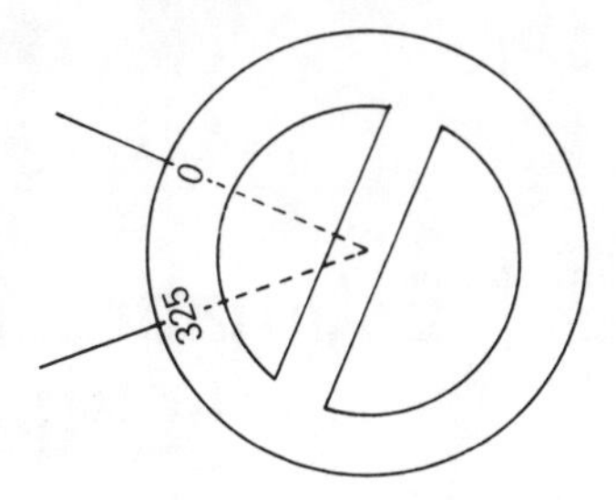

Fig. 24.8. Measuring a 35° angle, the hard way

tomatic. Students appropriately place the center of the protractor on the vertex of the angle, since this habit has been so well established in prior exercises. Because of the two scales appearing on most conventional protractors, students must, of course, be warned to read the intersections of the two sides of the angle with the arc from the same scale. Figure 24.9 shows some of the protractor positions and scale readings that students might use to measure a 52° angle correctly. Trying to find the "right" position and the "right" scale are no longer as essential as in the usual instructional sequence, since students are measuring an arc. And the arc is something that can be seen—it is a piece of the protractor's circumference. They are not just finding an abstract number to attach to *one* of the two rays of the angle. Thus students have been brought to the point of easy and meaningful use of the protractor.

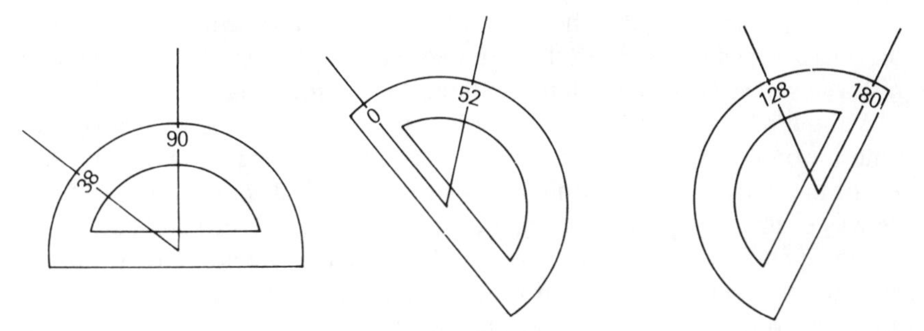

Fig. 24.9. Measuring a 52° angle

Continued emphasis on estimation, after the protractor has been introduced, is essential to maintaining the skill acquired in the earlier activities. Students should be encouraged to estimate angle measures before using their protractors, and oral and written exercises asking only for an estimate, like those in the introductory lessons, should be a component of further work.

25

Estimating the Size of "Little Animals"

James H. Wandersee

ANTONY van Leeuwenhoek (lay′ ven hōōk) was a seventeenth-century biologist who was truly a master microscopist. Having built more than two hundred microscopes himself, he fully understood the research possibilities of the newly invented instrument. His fascination with the tiny world behind his lens, coupled with his limitless patience and carefully written observations, led him to make many important biological discoveries.

With his homemade microscopes, Leeuwenhoek saw what no other scientist of his day could see. He even discovered bacteria and protozoa, affectionately calling them his "little animals." Visitors came from all over the world (for example, Peter the Great of Russia) just to peer through one of his microscopes at his carefully mounted biological specimens. No wonder historians of science called him the father of microbiology.

Although he had little formal training in mathematics, he was, at various times, a bookkeeper, cloth merchant, cashier, wine gauger, and surveyor. All of these occupations required some mathematical competence. Indeed, he was examined by the mathematician Genesius Baen in 1669 and found legally qualified to be a surveyor in his native Holland. This probably explains his ability to perform arithmetic calculations regarding the size, area, and volume of the microbes he described.

One problem Leeuwenhoek faced was measuring the *size* of those "little animals" that he observed. Today the measurement of small objects with a microscope is common practice and is relatively easy. Stage micrometers and eyepiece graticules simplify the job. Yet, in Leeuwenhoek's time, it was extremely difficult to measure small objects. Their size could only be estimated by reference to other objects of known dimension in the same field of view.

Here's how Leeuwenhoek attacked the problem. (See fig. 25.1.) Using the "inch" (of his country and time) as his standard unit, he determined that

30 coarse (or 100 fine) sand grains laid end to end equaled about an inch. Placing a referent coarse grain of sand in the same field as the "little animals," he noted that one microbe had a diameter of about one-twentieth the sand grain's diameter. Another microbe was about one-fifth the diameter of the first. An even smaller "animal" was about one-fifth the diameter of the second. From these estimates he was able to calculate the volume of each object and assert that a billion of the smallest animals would fit inside a volume like that of a coarse grain taken from fine scouring sand.

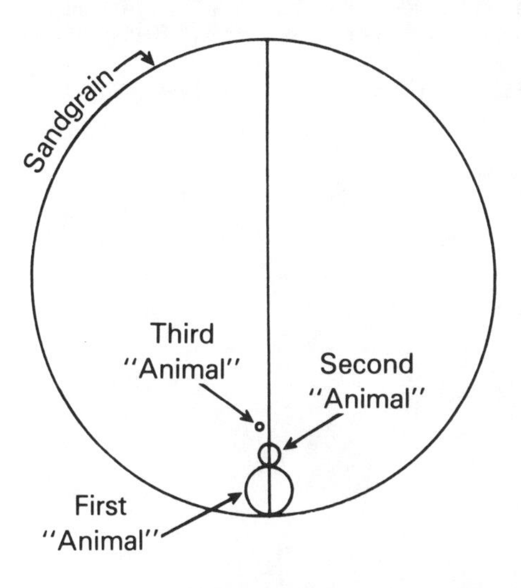

Fig. 25.1. Leeuwenhoek's method of estimation

It is interesting to note that Leeuwenhoek used the formula for the volume of a cube rather than a sphere to determine the size of his little animals, even though they were nearly spherical in shape. Although scientists today might gasp at such an error, it must be pointed out that Leeuwenhoek was the *first person* to measure anything using a microscope. His crude approximations helped biology become a quantitative discipline. Before such calculations, no one would have even dreamed that there were more bacteria in one person's mouth than there were citizens in the United Netherlands!

The descriptive and quantitative letters that Leeuwenhoek wrote to the members of the Royal Society of London to report the approximate size of the organisms he had discovered allow modern microbiologists to understand and verify his work today.

As the first micrometrist, Leeuwenhoek measured what he saw behind his lens using nearly constant biological "metersticks," such as sand grains, human red blood cells, vinegar eels, millet seeds, and the eye of a louse. The approximations he made helped him communicate the findings of microbiology to an eager public and opened a new frontier—inner space.

BIBLIOGRAPHY

Bulloch, William. *The History of Bacteriology.* London: Oxford University Press, 1938.

Dobell, Clifford. *Antony van Leeuwenhoek and His "Little Animals."* New York: Dover Publications, 1932.

Schierbeek, A. *Measuring the Invisible World.* London: Abelard-Schuman, 1959.

Wandersee, James H. "Antony van Leeuwenhoek: A Voice from the Past." Paper presented at the annual convention of the National Association of Biology Teachers, 24 October 1980, Boston.

———. "A Letter from Leeuwenhoek." *American Biology Teacher* 43 (November 1981): 450–51.

Note. Today, Antony van Leeuwenhoek's "little animals" are not considered animals at all. These microorganisms are classified in kingdoms Monera and Protista. Leeuwenhoek's microscopes were completely different from the kind scientists use today. They had a single lens, were hand-held vertically, and were smaller than an audiocassette tape.

Benchmark: Two Folk Methods for Estimating Distance

David R. Duncan
Bonnie H. Litwiller

IN REAL-WORLD situations it is frequently necessary to estimate lengths for practical purposes. Some methods of performing these estimations have been commonly taught in family or work settings rather than in schools and may thus be called folk methods. We shall describe two folk methods for estimating distances, the rod and the mile.

Farmers frequently make use of the linear measure called the *rod.* A rod is defined to be 16 1/2 feet in length. Its usefulness stems from its connection with the unit of area called the *acre*. By definition, one square mile (often called a *section*) contains 640 acres. A section is frequently subdivided into four squares of 1/2 mile each side. Such a square (called a *quarter-section* or simply a quarter) has an area of 1/4 square mile, or 160 acres. Since 1/2 mile is 2640 feet in length and since 2640/160 = 16.5, a strip of land 1/2 mile in length and 1 rod in width has an area of 1 acre.

A farmer with a quarter-section often wishes to subdivide this large field into strips having specific acre sizes. To do so, a common folk method is to "step off" the desired number of rods on one side of the quarter. To do this, five single steps are used to approximate 1 rod. The step must have a length of 16.5/5 = 3.3 feet—just over a yard (and, coincidentally, barely 2/10 of an inch over a meter).

To develop the correct step length requires a great deal of practice. This skill has been an important one for farmers to acquire, since land divisions, seed purchases, and yield determinations may depend on the accuracy of the farmer's steps.

Although the rod is an English unit unrelated to the metric system, it is interesting to note that the step to be used in this rod approximation process is 39.6 inches long—barely 0.2 of an inch longer than a meter.

Historically, the mile originated as the distance covered by a Roman soldier in 1000 double steps. This origin can be seen in the similarity between the words *mile* and *mille*—meaning 1000. Since 5280/2000 = 2.64, a single step used by a soldier was 2.64 feet in length, or approximately 2 feet 8 inches.

It should be noted that the step length necessary for a soldier to step off a mile is considerably shorter than that used by a farmer to step off a rod; the soldier's step is much easier to maintain over a long distance than the farmer's special-purpose long step.

A common use of an estimation is to find usable values in situations where precise values exist but are difficult to secure. Both of these folk methods are examples of this important use.

26

Evaluating Computational Estimation

Robert E. Reys

COMPUTATIONAL estimation is a basic skill. Yet despite the recent widespread support for giving more attention to it in school mathematics programs, much remains to be done. Evaluation of computational estimation, for example, continues to be one of the most neglected areas of test development. Over the years, testing programs have often omitted estimation completely or at best made very shallow efforts to assess estimation skills.

The purposes of this article are to (1) discuss the few inadequate efforts that have been made to test estimation and identify some reasons why more has not been done and (2) provide some guidelines for improving the testing of estimation. It is hoped that this discussion will stimulate an exchange of ideas, thereby promoting further progress in the testing of computational estimation.

EFFORTS MADE TO TEST ESTIMATON

Think about your own background and experience. How are your computational estimation skills? Were they ever tested in school? Probably not, yet most of us are faced with many situations in which we need to apply computational estimation skills in our daily lives:

Example 1: "How much tip should I leave for a $12.86 meal?"

Example 2: "Which is the better buy, 650 g for $1.28 or 850 g for $1.95?"

Example 3: "Is $10 enough to buy three items costing $0.89, $4.95, and $2.19?"

These examples illustrate several different types of computational estimation problems. Example 1 requires that a specific amount of money be left

The author gratefully acknowledges the support provided by the Research Council of the University of Missouri—Columbia in the preparation of this manuscript.

for a tip, but the amount we leave is probably an "estimate," that is, about 15 percent of $12.86. Example 2 suggests a direct comparison to determine the better buy. This decision may be made quickly by recognizing that one price is below .2 cents a gram and the other is more than .2 cents a gram. In example 3, the upper bound of $10 illustrates the importance of a reference point in estimation. Often real-world constraints in problems serve as reference points.

There are, of course, many other examples of applications of computational estimation (Reys and Reys 1983). In fact, a recent survey of non-school uses of mathematics reported that in more than 80 percent of the situations surveyed an estimate of the answer was considered sufficient. Estimation is clearly in widespread use. Yet all standardized tests focus on paper-and-pencil computation, giving little or no attention to computational estimation. This is a classic example of school testing programs either not recognizing or simply ignoring life skills in their assessment efforts.

Certainly computational estimation is difficult to assess. It has been said that estimation as a process is difficult to measure and "the criteria for good estimates are often ambiguous" (Carpenter et al. 1976, 299; Reys and Bestgen 1981; Reys 1983). The psychometric difficulties of testing computational estimation may discourage many bona fide assessment efforts from addressing it at all. Consider, for example, the statewide assessment in Missouri, the Basic Essential Skills Test (BEST), which is given to every student in the eighth grade or later. The BEST explicitly states that estimation is viewed as a basic mathematical skill, yet estimation was not included when a statewide test of "basic" mathematical skills was constructed. It is interesting that the state education agency would omit estimation from its basic skills test and yet advise school districts that the responsibility for assessing student performance in estimation rests with the local school.

Why is computational estimation so difficult to assess? Consider, for example, the following potential test question: What answer would be

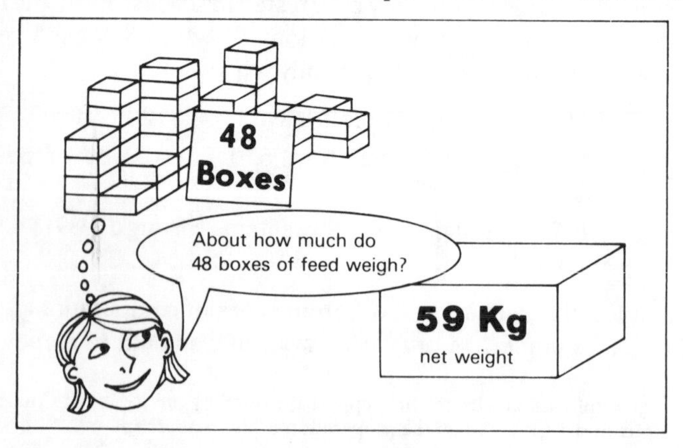

indicative of good estimation ability? Consider also the following scenarios for three hypothetical students:

> Rex computed 48 × 59 on his paper and found the product to be 2832. He then rounded 2832 to 2800 and reported his "estimate" to be 2800.
>
> Pat rounded 48 × 59 to 50 × 60 and mentally found the product to be 3000. She reported her "estimate' to be 3000.
>
> Julio rounded 48 × 59 to 50 × 60 and mentally found the product to be 3000. He thought this would be too high, so he adjusted his answer and reported his "estimate" to be 2800.

Each of these answers is an acceptable estimate. Pat and Julio both used the technique of rounding, but Julio's adjustment process went beyond the initial estimate of 3000 to produce a refined estimate of 2800. Although Rex and Julio produced the same estimate, their processes were very different. Whereas Julio used several important estimation techniques, Rex never estimated. Rather, he applied a written computation algorithm and then rounded his exact answer to satisfy the directions.

When asked to estimate, students often try to work a problem quickly with pencil and paper and then round their answer to reflect an "estimate" (Bestgen et al. 1980, 11). This procedure is clearly not computational estimation, but the tendency to use it on paper-and-pencil assessments is a major reason why estimation skills are difficult to assess. This paradox—getting the "right answer for the wrong reason"—has discouraged the authors of many standardized tests from including questions related to computational estimation.

Standardized achievement tests are widely used today to evaluate students and school programs. Writers and publishers of achievement tests claim their tests mirror mathematics programs, that is, they claim to test what is taught in most programs today. In an effort to determine the attention given to computational estimation, the most widely used standardized mathematics achievement tests for grades 4–8 were reviewed. The following conditions were revealed:

- The *California Achievement Test* (1977) contained no computational estimation items in four levels. On one level, one out of forty-five exercises on Concepts and Applications involved estimation.
- The *Iowa Test of Basic Skills* (1978) contained no computational estimation items in two levels. On the other three levels, one out of forty-five exercises on Computation involved estimation.
- The *Metropolitan Achievement Tests* (1978) contained no exercises in either the Intermediate or Advanced levels that involved computational estimation.
- The *SRA Achievement Series* (1978) included no computational estima-

tion items in three grades. In one level, one out of forty computation exercises required estimation. In another level, two of the forty computation exercises used the word *estimated,* but consider the following paraphrase of one of those exercises:

> The team practiced 2 times a week for 3 weeks. Each practice lasted 1 1/2 hours. To estimate the number of hours the team practiced, you should find the answer to
>
> A. (2 + 3) × 1 1/2 C. 2 × 3 × 1 1/2
>
> B. 2 × 3 + 1 1/2 D. (2 × 1 1/2) − 3

Although the word *estimate* was used in this exercise, no estimation is required. The word *estimate* is not used as a mathematical process but rather as a synonym for *determine* or *find* or *compute.*

- The *Stanford Test of Academic Skills* (1983) contained no computational estimation items on the Intermediate 1 level. On both the Intermediate 2 and Advanced levels, three out of forty-four exercises from the Computation subtest involved estimation.

This review showed that the coverage of computational estimation varies among achievement tests but is, at best, sketchy and inadequate. At some grade levels estimation was not even included, and on no test was computational estimation given the attention that has been recommended by professional organizations and commissions (National Council of Teachers of Mathematics 1980; Conference Board of the Mathematical Sciences 1983).

The testing and evaluation of students in mathematics must include the measurement of computational estimation skills. Indeed, one major criterion for selecting a standardized test of mathematics should be the breadth and depth with which it assesses estimation skills.

Validly measuring computational estimation skills is clearly a challenge, but it can and must be done. And some progress has been made. For example, over twenty years ago the National Longitudinal Study of Mathematical Abilities (NLSMA) recognized the importance of testing computational estimation and included it in its test batteries. Figure 26.1 shows one exercise.

If there are 5,280 feet in one mile, how many feet are there in 7 miles?

A. 35
B. 350
C. 3,500
D. 35,500
E. 350,000

Fig. 26.1 A sixth-grade exercise from NLSMA

It is interesting that NLSMA prefaced its entire cluster of estimation items with a question mark to alert readers that caution should be used in interpreting the results, since the test developers were not certain about the validity of these exercises. In fact, NLSMA found evidence that many students actually performed the calculations (Wilson, Cahen, and Begle 1968). It is to the credit of NLSMA researchers that they not only attempted to assess estimation but also were honest in sharing their own concerns about their ability to obtain a valid measure.

Another large-scale mathematics assessment, which included computational estimation, was conducted by the National Assessment of Educational Progress (NAEP). Consider, for example, the exercise in figure 26.2, which comes from the first mathematics assessment (1972–73).

John has 382 stamps in his stamp collection. Greg has 224, Pete has 310, and Bob has 175. The number of stamps the boys have altogether is CLOSEST to which of the following numbers?

900
1000
1100 (correct response)
1200
No response or "I don't know"

Fig. 26.2

Do you have any reaction to this exercise? One valid criticism is that the closeness of the choices may force too much precision, thus discouraging the use of estimation in favor of exact computation. Despite such shortcomings, poor performance on items like this one is the basis for the comment that estimation is "one of the most neglected skills in the mathematics curriculum" (Carpenter et al. 1976, 296). Results from the second and third mathematics assessments have also shown low performance on computational estimation (Carpenter et al. 1981; Carpenter et al. 1983). In fact, in the third NAEP mathematics assessment, performance on computational estimation items was significantly lower than on the parallel exact computation items. This is illustrated in figures 26.3 and 26.4.

Compute: $\begin{array}{r} 526 \\ \times\ 4 \\ \hline \end{array}$

Answers	Number of 13-year-olds giving that answer
522	0
530	1

2004	4
2084	2
2094	1
2104	89
IDK	2
No response	1

Fig. 26.3. Parallel NAEP exercise from the third assessment requiring exact computation

Answers	Number of 13-year-olds giving that answer
2000 g	23
2100 g	33
2400 g	21
2000 g	13
IDK	8
No response	2

Fig. 26.4. Parallel NAEP exercise from the third assessment requiring computational estimation

An examination of the results shows that almost 90 percent of the 13-year-olds computed an exact answer correctly compared to only 33 percent of the same age group that did the estimation correctly. All the differences between exact computation and estimation were not this great, *but the performance was always higher on exact computation.* The dramatic shifts, usually about 20 to 30 percent on these parallel exercises, suggests that computational skills and computational estimation are very different.

In order to increase validity, NAEP has made efforts to control the time allotted to each computational estimation item and to provide different types of problem situations. It is also significant that the breadth and scope of items addressing computational estimation have increased with each assessment. Not only have whole numbers, fractions, decimals, and percents been included, but NAEP has also included a variety of mathematical operations in its estimation exercises.

Inclusion of computational estimation by NAEP, even though just a first step, is encouraging. Furthermore, the consistently poor performance on estimation items for all age groups has helped stimulate discussion about teaching estimation. It is hoped that these factors, in addition to increased attention to instruction in computational estimation, will encourage producers of standardized achievement tests to include more computational estimation in their tests while simultaneously decreasing the emphasis on traditional paper-and-pencil computation. Such a turn of events would give each type of computation the attention it deserves.

SOME GUIDELINES FOR TESTING ESTIMATION

Techniques to assess computational estimation are still in their embryonic stages of development. However, in addition to the large-scale assessment efforts by NLSMA and NAEP, a number of smaller research projects have been productive (Driscoll 1981 and 1982; Paull 1971; Reys et al. 1980; Reys et al. 1984; Rubenstein 1982; Schoen et al. 1982). More research needs to be done, but some guidelines for testing estimation have emerged from the efforts to date. These guidelines will be discussed in the following four sections.

Timing

The time allocated to make estimates should be carefully controlled, with the amount of time depending on three important factors: (*a*) the operations—division usually takes longer than addition; (*b*) the format—$47\overline{)8526}$ takes less time than 8526 divided by 47; and (*c*) the complexity of the numbers—216 × 859 takes more time than 21 × 85. Even though these factors may increase the amount of time that should be allowed, the testing time for estimation must nevertheless be carefully controlled. If too much time is allowed, the test may only be a measure of paper-and-pencil computation. If too little time is allowed, only wild guessing may occur. Somewhere between these two extremes will be an optimal time that can often be determined by experimenting with students. Research related to timing confirms that allowing just five seconds more for each question produces dramatic differences in performance as well as in the processes used. Although additional research is needed, it is my belief that to ensure that students estimate and don't compute, it is better to give too little time than too much.

Furthermore, it is recommended that estimation items be administered and timed individually. An overhead projector, slide projector, or microcomputer are all effective means to present items one at a time while maintaining control over the amount of time for each item. If test booklets are used instead, it is essential to control the amount of time available to complete each page. The testing directions must make it clear that written algorithms are not to be used. Even then it is impossible to monitor an entire class to ensure that the directions are followed. For example, both NLSMA and NAEP controlled the time on each page, yet both reported that some students computed exact answers. Such students just plod through the test doing traditional paper-and-pencil computation, and thus their scores are not an indication of their computational estimation skills.

Regardless of how the test is administered, the total test time should probably not exceed ten to fifteen minutes. The timing constraint on each item guarantees a rapid pace that requires high levels of attention and concentration. These factors produce fatigue very quickly, so keeping the entire test time short is essential. Exactly how short depends on other factors, such as the formats for the answers.

Formats for Answers

Several different answer formats for testing computational estimation have been used. Although open-ended questions are recommended in general, various formats, along with some advantages and disadvantages of each, will be discussed. Consider this problem:

Example 4: If you deliver 95 newspapers a day, about how many papers will you deliver in a year?

Five approaches to testing estimation might be used:

1. *Open-ended.* A particular value or estimate must be constructed. For example, a person might think, "95 is about 100, and 100 times 365 days is 36 500." Another person may adjust the estimate by thinking, "Since 95 is less than 100, I need to subtract something—so my estimate is 35 000." Open-ended questions provide no clues about the answer, and their nature guarantees that a range of estimates will be produced. Thus, in order to score an open-ended question, the teacher must establish an "acceptable interval" for students' estimates. An efficient way to determine the acceptable interval involves taking the correct answer plus or minus a certain percent. This is an easy rule to apply, but it has some obvious shortcomings. For example, estimate this sum:

$$98 + 295 + 485$$

An estimate of more than 900 seems unacceptable by any standard, yet if a 10 percent error is allowed for this problem, the acceptable range (790 to 966) would exceed 900.

To avoid such inconsistencies due to mechanical rules, it is recommended that possible solution strategies be used to determine the acceptable intervals. That is, the most meaningful way to establish an interval is to identify strategies that would be appropriate for solving the problem and determine the upper and lower values of their resulting responses. The least lower and greatest upper values can then be used as endpoints for the acceptable interval. For example, 100 × 365 suggests an upper bound of 36 500 for example 4. The lower bound results from some adjustments and might be 33 000. Although establishing acceptable intervals in this way takes a great deal of time, it is consistent with the notion that there are several good estimates—arrived at by using different, but appropriate, strategies—for any open-ended estimation problem. Unfortunately, the need for hand scoring precludes the use of open-ended test items on most currently available standardized tests. Other formats, such as the ones that follow, though not as open-ended, can be used on standardized tests.

2. *Intervals.* A pseudo number line is provided and some predetermined values are placed along the continuum. For example, again refer to the problem posed in example 4, where some intervals might be shown as follows:

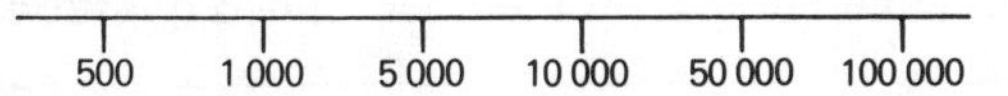

This approach has been used in large-scale testing (Herbert and Dougherty 1983). It requires the student to place an estimate along the number line and darken the appropriate cell on the test answer sheet. The intervals provide boundaries for the estimate, eliminating the need to establish them in advance, as in open-ended items. However, the labeling of the continuum must be done with care.

In this approach students choose from among intervals; hence, the items can be machine scored. On the negative side, the changes in scale on the continuum (i.e., showing the same distance on the number line between 100 and 500 as between 10 000 and 50 000) can be confusing.

3. *Multiple Choice.* A choice must be made among given values. The alternatives are constructed to provide choices that reflect frequent errors. For example, these choices might be provided for example 4:

A. 90
B. 100
C. 350
D. 450
E. 35 000

These values may seem random, but closer examination suggests that each of them could be selected by the use of a systematic, albeit incorrect, strategy. For example, 90 or 100 would result if only the number 95 is considered in the problem. If only the number of days in a year is considered, 350 might be chosen, and 450 would be a plausible alternative for anyone adding 95 to the number of days in a year.

Constructing attractive foils, such as these, in a multiple choice test is challenging. A recommended approach is to examine the errors students make on open-ended questions.

Multiple choice items, to their advantage, are machine scorable. However, some research suggests that students often use a different approach to multiple choice than to open-ended estimation questions (Reys et al. 1980). For example, when given a multiple choice estimation item, students often do not produce a direct estimate but rather consider each choice individually and decide whether to accept or reject it.

There is also evidence that the percentage of correct responses will differ for parallel questions, one in multiple choice and the other in open-ended form. For example, consider the exercise in figure 26.5, given to two similar groups of ninth graders (Reys et al. 1980). Each form was presented with a slide projector for twelve seconds. On the multiple choice item, 37 percent of the students correctly chose 0.3, whereas only 28 percent gave a response within the acceptable interval on the open-ended question.

Multiple Choice	Open-ended
$73\overline{)22}$	$73\overline{)22}$
a. 0	
b. 0.3	(Acceptable interval 0.25–0.40)
c. 0.5	
d. 1	
e. 3	
f. Don't know	

Fig. 26.5

Thus, although multiple choice estimation items are machine scorable, they have a number of limitations. A format that combines aspects of the multiple choice and intervals formats is considered next.

4. *Order of Magnitude.* A choice must be made among estimates of different orders of magnitude. This is an integration of the interval and multiple choice formats. Consider again example 4, with the following choices provided:

A. 360 papers   C. 360 000 papers
B. 3 600 papers   D. None of these

Although the order-of-magnitude format can be used with any estimation problem, it is particularly appropriate for multiplication and division estimation, since these operations are likely to lead to order-of-magnitude errors. For estimation with other operations, other formats in our continuing list should be used.

5. *Reference Number.* A decision must be made about whether an exact answer is above or below a given reference number. Return once again to example 4 and consider 36 000 as the reference number. The student's task is to decide whether 36 000 is larger or smaller than the number of newspapers that you would deliver in a year.

Reference number problems abound in everyday living. For example:

- My salary is $22 950 a year. Is that $2 000 a month?
- There are 458 students. 51 of them are absent. Are 20% of them absent?

Reference number problems are very closely related to the task of determining whether an answer is reasonable. This relationship is clear if one thinks of the reference number as the answer whose reasonableness is to be tested. To see this, compare the following example to the analagous estimation problem with 400 as reference number.

- The last time I checked the gasoline mileage on my car, I got 21 miles a gallon. Would it be reasonable for me to expect to drive 400 miles on 15 gallons?

Since one of two choices must be made, items in the reference number format lend themselves well to machine scoring. The trade-off, of course, is that students can guess with 50 percent chance of success. However, items in the reference number format can be completed quickly, allowing for many problems on a test. Thus, the negative effect that guessing has on the reliability and validity of the estimation test can be offset by lengthening the test.

Clearly, each of these five formats for estimation test items has strengths and weaknesses. Whatever format is chosen, decisions concerning the numbers to use in the questions must also be made. Guidelines for making those decisions follow.

Numbers in Estimation Items

The numbers used in the test questions should be complex enough to encourage and reward estimation. Consider these two problems:

ESTIMATE

A. 425
+203

B. 4 358 529
+2 037 468

It is quicker and easier in problem A to mentally compute the exact answer than to estimate. On problem B it is much easier to make an estimate of "over six million" than to find the exact answer. Problem A tests mental computation, whereas problem B is a bona fide estimation item. In an estimation item, it should be clear to respondents from the numbers in the problem that there is not enough time to compute an answer—and that even if there were, it would be a lengthy, laborious exercise.

An estimation test should contain only problems that require estimation and avoid those that lend themselves to traditional paper-and-pencil algorithms or mental arithmetic. When students encounter situations where they can produce good estimates much faster than exact answers, their respect for the power of estimation increases. For example:

ESTIMATE

7/8 + 1 17/18 + 8/9

This becomes very messy using exact computation, but it is both quick and easy to produce an estimate of "almost four" by recognizing that each of these fractional parts is a little less than one. Obtaining an acceptable estimate for this example requires, as estimation items should, good number sense; in particular, a recognition of the size of the fractions.

Question Context

Both problems involving numbers only and problems in an applied context should be included on an estimation test. The research data shown in table 26.1 suggest that performance can change dramatically when the context of a problem is altered (Reys et al. 1980).

TABLE 26.1
Percent Correct on Parallel Computation and Application Questions

Problem	Grades 7–8 $N = 431$	9–10 $N = 359$	11–12 $N = 291$	Adults $N = 106$
Computation: 1 1/2 × 1.67 (acceptable interval 2–3)	23	45	40	60
Application: I need 1 1/2 yards of material. It costs $1.67 a yard. About how much will it cost? (acceptable interval $2–$3)	54	75	71	81

In this example, the numerical values and the operations required in the two problems were identical, yet respondents performed much better (at least 20 percent) when the problem was presented in an applied context. Other contexts can be constructed in which performance drops. Such research simply suggests that the context does indeed influence estimation performance, so a balance of different contexts is necessary to get the most meaningful measure of computational estimation skills.

CLOSING THOUGHTS

Computational estimation is an important part of the elementary and secondary school mathematics program. Just as no forward-looking mathematics program can ignore problem solving, neither can it ignore computational estimation. Since pressures are increasing to teach estimation, assessment procedures must be developed to provide a profile of individual performance and to monitor student progress over the years.

For decades, standardized achievement tests have included computational skills as a common core of their assessment packages. They have consistently tested addition, subtraction, multiplication, and division for whole numbers, decimals, and fractions. A similar template must now be constructed to measure computational estimation skills. Significant changes will occur only when schools demand that computational estimation be seriously addressed within major achievement tests. It is therefore recommended that a major criterion for selecting a standardized mathematics achievement test be the quality and scope of its assessment of computational estimation.

Despite the psychometric difficulties associated with testing computational estimation, progress is being made. It is hoped that the challenge of creating effective new approaches to testing computational estimation will motivate test companies to break away from the existing shackles of tradition and explore some new alternatives. Clearly many approaches exist for testing estimation, and the increasing availability of microcomputers holds exciting potential for assessment. Although research has provided some valuable insight into this testing question, much more research is needed to provide both reliable and valid measures of estimation ability.

The recommendations, practical suggestions, and guidelines offered in this article come from research and classroom experience and are in no way intended to be exhaustive. They are offered in the hope of stimulating the development of new and creative ways of measuring computational estimation skills.

REFERENCES

Bestgen, Barbara J., Robert E. Reys, James F. Rybolt, and J. Wendell Wyatt. "Effectiveness of Systematic Instruction on Attitudes and Computational Estimation Skills of Preservice Elementary Teachers." *Journal for Research in Mathematics Education* 11 (March 1980): 124–36.

Carpenter, Thomas P., Terrence G. Coburn, Robert E. Reys, and James W. Wilson. "Notes from National Assessment: Estimation." *Arithmetic Teacher* 23 (April 1976): 296–302.

Carpenter, Thomas P., Mary K. Corbitt, Henry S. Kepner, Jr., Mary M. Lindquist, and Robert E. Reys. *Results from the Second Mathematics Assessment of the National Assessment of Educational Progress.* Reston, Va.: National Council of Teachers of Mathematics, 1981.

Carpenter, Thomas P., Mary M. Lindquist, Westina Matthews, and Edward A. Silver. "Results of the Third NAEP Mathematics Assessment: Secondary School." *Mathematics Teacher* 76 (December 1983): 652–59.

Conference Board of the Mathematical Sciences. *The Mathematical Sciences Curriculum K–12: What Is Still Fundamental and What Is Not.* Washington, D.C.: National Science Foundation, 1983.

Driscoll, Mark J. *Research within Reach: Elementary School Mathematics.* Reston, Va.: National Council of Teachers of Mathematics, 1981.

———. *Research within Reach: Secondary School Mathematics.* Reston, Va.: National Council of Teachers of Mathematics, 1982.

Herbert, M., and K. Dougherty. *Summaries of Evaluation Reports of the Comprehensive School Mathematics Program.* St. Louis: MCREL, 1983.

National Council of Teachers of Mathematics. *An Agenda for Action: Recommendations for School Mathematics of the 1980s.* Reston, Va.: The Council, 1980.

Paull, Duane R. "The Ability to Estimate in Mathematics." Doctoral dissertation, Columbia University, 1971.

Reys, Barbara J., and Robert E. Reys. *Guide to Using Estimation Skills and Strategies (GUESS).* Palo Alto, Calif.: Dale Seymour Publications, 1983.

Reys, Robert E. "Mental Computation and Estimation: Past, Present, and Future." *Elementary School Journal* 84 (May 1984): 547–57.

Reys, Robert E., and Barbara J. Bestgen. "Teaching and Assessing Computational Estimation Skills." *Elementary School Journal* 82 (1981): 117–27.

Reys, Robert E., Barbara J. Bestgen, James F. Rybolt, and J. Wendell Wyatt. *Identification and Characterization of Computational Estimation Processes Used by In-School Pupils and Out of School Adults.* Final Report, Contract No. NIE-G-79-0088. Washington, D.C.: National Institute of Education, 1980.

Reys, Robert E., Paul R. Trafton, Barbara Reys, and J. Zawojewski. *Developing Computational Estimation Materials for the Middle Grades.* Final Report, Grant No. NSF81-13601. Washington, D.C.: National Science Foundation, 1984.

Rubenstein, Rheta N. "Mathematical Variables Related to Computational Estimation." Doctoral dissertation, Wayne State University, 1982.

Schoen, Harold L., Charles D. Friesen, Jocelyn A. Jarrett, and Tonya D. Urbatsch. "Instruction in Estimating Solutions of Whole Number Computations." *Journal for Research in Mathematics Education* 12 (May 1981): 165–78.

Wilson, James W., L. S. Cahen, and Edward G. Begle. *NSLMA Reports.* No. 4. Palo Alto, Calif.: Stanford University, 1968.

27

A Summary of Research on Teaching and Learning Estimation

Sidney E. Benton

WITH the current emphasis on estimation skills reflected in this yearbook, one would expect that teachers are now doing more to teach these skills. One might also expect that there would be numerous new research studies on estimation. However, Siegel, Goldsmith, and Madson (1982) reported that "although the literature of mathematics education has argued repeatedly for the importance of estimation in elementary and intermediate mathematics curricula, there has been surprisingly little emphasis on research or curriculum development of estimation skills" (pp. 211–12). In the annual listings of "Research on Mathematics Education" for 1979–83, Suydam (1982, 1983, 1984) and Suydam and Weaver (1980, 1981) reported 74 research summaries, 1058 published journal reports, and 1711 dissertations. Of these, only 12 (less than one-half of 1 percent) deal with the teaching and learning of estimation in any substantive way. According to Reys, Rybolt, Bestgen, and Wyatt (1982), "Research studies of computational estimation are rare, because such skills are difficult to test" (p. 184). Some of the difficulties were pointed out by Reys and Bestgen (1981): "The very nature of computational estimation makes a formal assessment difficult. Unless the time allowed for a problem is carefully controlled, a complete computation rather than an estimation may actually be performed" (p. 123). R. E. Reys discusses the testing of estimation ability in more detail in article 26.

It is important for the mathematics educator to be aware of the findings that do exist, however. This article reviews the research and groups it into three major areas: estimation ability of students and teachers, strategies and characteristics of estimators, and experimental studies on estimation.

STUDENTS' AND TEACHERS' ESTIMATION ABILITIES

A number of research studies indicate that performance on estimation tests is at best disappointing. Faulk (1962) investigated the ability of 52 fifth-grade students from two classrooms to estimate answers to this verbal problem: "If a gallon of gasoline costs 26¢, how much will 15 gallons cost?" (p. 437). Only one-half of the answers were in the "correct" range of $3.00 to $4.99.

Others (Carpenter et al. 1976; Reys and Bestgen 1981; and Reys et al. 1982) reported low performance on estimation items from the National Assessment of Educational Progress (NAEP). For example, only 54 percent of the 17-year-olds could adequately estimate to the nearest million dollars the sum of four addends: $11 954 164 + $1 126 005 + $4 170 522 + $750 572 (Carpenter et al. 1976).

Students had even more difficulty estimating answers to exercises involving fractions or decimals. Reys and Bestgen (1981) reported that when students were asked to estimate the sum of three decimals, "such as 295.0 + 865.2 + 1.583, more than 70 percent of the 13-year-olds and 50 percent of the 17-year-olds chose either 10,000 or 100,000 as the sum" (p. 119). Carpenter and his colleagues (1980) pointed out that students' inability to estimate fraction addition probably had more to do with not understanding the concept of fractions than with a lack of estimation ability.

Several studies have investigated students' perceptions of weight, distance, height, area, and temperature. Again, results indicate on the whole that students give inaccurate estimates of these measures.

Corle (1960) investigated the quantitative concepts of 39 fifth graders and 108 sixth graders in Pennsylvania. The students were asked to estimate ten quantities, including the weight of a block of wood, the length of a piece of rope, the circumference of a basketball, the temperature of the room, and the amount of water in a pail. Students were individually interviewed outside the classroom. The fifth graders gave estimates that averaged almost six times the actual measurements, whereas the sixth graders' estimates averaged one and one-half times the actual measurements. Corle concluded that estimating can be learned in a school year and maturation is a factor in the development of the skill. He further observed that boys estimate such quantities more accurately than girls do.

Corle (1963) also investigated the estimation ability of adults. Subjects were 368 elementary school teachers and 96 college juniors majoring in elementary education, who were asked to estimate various quantities. Not surprisingly, these subjects' estimates seemed to be more accurate than those of fifth and sixth graders (p. 352). However, their average error was about 60 percent, with the teachers appearing to be more accurate than the college students.

Paull (1972) analyzed the ability of 196 eleventh graders to estimate length, area, and answers to numerical computations. He found the students inconsistent in their ability to estimate answers across the various types of problems. Carpenter and his colleagues (1976) stated, "This would imply that students need a wide variety of broadly based estimation activities"

(p. 299). Paull also reported that there were positive correlations between both mathematical and verbal ability and the ability to do computational estimation but not the ability to estimate length. He also reported no significant sex differences in ability to estimate answers to the various types of problems. However, one should be cautious in making generalizations from the Paull study, since all the subjects were enrolled in college preparatory mathematics classes.

In addition to Corle's 1963 study, several studies have examined the ability of college students to estimate. Swan (1970) and Swan and Jones (1971) reported a distressing inability of college students to estimate various quantities. Estimates by 200 college juniors of the weight of a 13-pound block of wood ranged from 3 to 45 pounds; the height of a 60-foot tower, from 20 to 200 feet; and the width of a 24-foot building, from 10 to 38 feet (Swan 1970). Metric estimates were even worse, with most of the students unable to make any estimates at all in meters, kilograms, or degrees Celsius.

In their study of 392 students preparing to become teachers, Swan and Jones (1971) concluded that the students had invalid perceptions of distance, weight, height, area, and temperature in both English and metric units. The majority of the estimates in English units, with the exception of height and temperature, were in error by more than 25 percent. Estimates using the metric units were again less accurate than those made with English units.

Swan and Jones reported in 1980 the results of their study to determine if a change had occurred in measurement estimation ability between 1971 and 1977. In addition to college students, subjects included students from grades 4–12. As in 1971, schoolchildren and adults had inaccurate perceptions. In only a few of the groups did as many as one-half of the subjects estimate with reasonable accuracy. As in previous studies perceptions were more accurate when expressed in English units; however, there was significant improvement from 1971 to 1977 in the estimates of the students using metric units, especially in elementary and junior high school.

Perhaps one reason students have not performed well on estimation tests is that teachers do not feel competent to teach estimation. In an opinionnaire given to 1108 practicing teachers by Stahlecker (1979), respondents indicated that they felt least adequately prepared in the areas of measurement, estimation, and the metric system.

These ability studies show, on the whole, across all age levels and for both computational and measurement estimation a disappointing inability to estimate.

STRATEGIES AND CHARACTERISTICS OF ESTIMATORS

Few researchers have examined strategies or processes used by estimators; however, of those who have, many argue that the strategies they observed should be taught in mathematics classrooms. Several also sought to determine whether variables such as intelligence, sex, and grade level were related to estimation ability.

One such study (Reys et al. 1982) used the Assessing Computational Estimation (ACE) test to select 59 "good" estimators from a sample population of 1200, which included students in grades 7–12 and selected adults. These good estimators were then interviewed to determine the strategies and processes they used to solve estimation problems. All had a quick and accurate recall of the basic facts, an understanding of place value to determine accurate results, facility with mental computation involving rounding numbers, skill at rounding to multiples of 10, and tolerance for error in the estimation process. In addition, a majority understood and could apply basic number properties and used one or more of these strategies: *reformulation* (rounding to manageable or compatible numbers), *translation* (restructuring the problem to a more mentally manageable form), and *compensation* (adjusting the numerals to compensate for reformulation or translation). Twenty to thirty percent of those interviewed used a variety of strategies, were confident about their estimation ability, were efficient at mental computation with all types of numbers, and used intermediate compensation processes.

The authors cautioned that the results from the adult sample should not be generalized to the entire adult population, since nearly all in the sample were college graduates and successful in their chosen occupations. The authors recommended that the three key processes identified in the report—reformulation, translation, and compensation—be further researched.

In another study, Hildreth (1981, 1983) identified the strategies used by 24 fifth graders, 24 seventh graders, and 24 college students in estimating length and area. He also compared strategy use with estimation ability and mathematical ability. He reported that strategies used by the fifth and seventh graders were not "qualitatively" different from those used by the college students. In the study, estimation ability and strategy use were positively correlated to perceptual ability. They were also positively correlated to mathematical ability for the college students but not for the others. There was a positive correlation between estimation ability and strategy use for all students. Finally, and somewhat surprisingly, no grade level or sex differences were reported on estimation ability or strategy use. Unfortunately, Hildreth (1983) used terms such as "positively correlated" and "correlated" without mentioning the size of the correlations. The reader should, it seems, view the reported relationships with caution, since in a study of this nature a correlation can be statistically significant yet too small to be useful for prediction purposes.

Of potential use to teachers are the nine appropriate estimation strategies and five inappropriate strategies described by Hildreth (1983, 50–52). The appropriate strategies for estimating length were *unit iteration, subdivision clues, prior knowledge, comparison of the target object with another object, chunking* (subdividing the length to be estimated into congruent segments), and *squeezing* ("making estimates that are a little less and a little more than the object" in order to narrow in on the measurement of the object). The appropriate strategies for estimating area were *repeated addition, length times width,* and *rearrangement.* Inappropriate strategies included *length plus width, count around* ("estimating the area of a rectangle using a nonstandard rectangular unit by estimating the length of the rectangle with the unit turned in one direction and width of the rectangle with the unit turned 90 degrees"), *centering* ("reporting the area of an object as being the estimated length of one of its dimensions"), and *wild guessing*.

In another study of estimation strategies, Levine (1981) also investigated the correlation between the number of estimation strategies and quantitative ability. Subjects were 89 college undergraduates who were not mathematics majors. Scores on the quantitative ability subtest of the School and College Ability Test and the researcher-constructed Test of Estimation Ability were obtained for each subject. Levine reported, "The data suggested that estimation is a difficult task for college students who are not mathematics majors, particularly for those of low quantitative ability" (p. 5013–A). She also reported a significant correlation of .55 between quantitative ability and the number of different estimation strategies used.

Levine further identified nine types of difficulties, misconceptions, and mistakes: incomplete process, loss of intermediate steps, incomplete strategy, incorrect meaning of an operation, inappropriately adjusting the result, incorrect algorithm, place-value errors in multiplication or division, rounding errors, and order-of-magnitude errors.

Finally, Rubenstein (1983) examined the computational estimation abilities of 309 eighth graders. She developed an estimation test designed to measure four types of computational estimation—*open-ended, reasonable versus unreasonable, reference number,* and *order of magnitude.* Subjects performed differently on three of these types of estimation, with open-ended being the most difficult. Males did better than females on total computational estimation and on order-of-magnitude estimation. There were no significant differences between the sexes on open-ended or reference-number estimation.

These few studies of estimation processes suggest that although there are a wide variety of viable estimation strategies, inappropriate strategies and misconceptions are likely to inhibit the ability of many people to estimate.

EXPERIMENTAL STUDIES

A small number of experimental research studies provide some evidence that instruction in estimation results in improved skill in solving estimation problems. The majority of these studies involve students in the middle grades; one deals with preservice teachers, and one with experienced teachers. No other experimental studies were found that involved college students, and none were found involving secondary students.

Since estimation has usually been introduced in the third or fourth grades, it is not surprising that very little research has been done in early childhood education. One exception is a study by Immers (1983) that included second graders. To determine the effect of estimation instruction on the linear estimation ability and strategy use of elementary students, Immers employed a pretest-posttest control group design. Forty-eight students in each grade from grades 2 through 5 (n = 192) from one school were randomly assigned to either an experimental group or a control group. A ten-day microcomputer-based instructional unit was prepared by the researcher. Immers reported that linear estimation ability correlated significantly with age for both treatment groups and that there were no significant sex differences. Instruction was most effective for the two weakest pretest areas—long and vertical estimation. Instruction benefited students in grades 3–5 but apparently not second graders. The experimental group performed significantly better on short and horizontal estimates and on the remaining subdivisions of the posttest than the control group. Interviews of subjects in a subsample revealed that unit iteration was the strategy most frequently used.

Another experimental study (Schoen et al. 1981) used subjects from two intact fourth-grade classes of 21 students each. The researchers developed five 45-minute estimation lessons designed to teach students "to learn to estimate the products of one-digit and two-digit, one-digit and three-digit, and two-digit and two-digit factors" (p. 166). In a separate part of the study, they also taught and tested fifth and sixth graders on similar estimation tasks. The researchers concluded that "estimation in whole number computation can be taught in a short period of time. Not only did fourth, fifth, and sixth graders become better estimators, they did so by adopting a valid estimation strategy. These skills and strategies were retained for at least a 3-week period . . ." (p. 176).

In addition to identifying strategies used in estimating length and area as described earlier in this article, Hildreth (1981, 1983) used the same 24 fifth graders and 24 seventh graders in an experimental study. Half the students in each grade were taught a "guess then check" method to estimate length and area in the metric system, and the other half were taught appropriate estimation strategies. Treatment consisted of five 40-minute lessons. Results indicated that the students who were taught strategies used more strategies on the posttest than those who were taught the "guess then check" method. Apparently both treatment groups had small significant improvement in strategy use and small significant improvement in estimation ability between the pretest and the posttest. Hildreth (1983) also reported a decrease in the number of inappropriate strategies, especially in the number of wild guesses. It is not clear in the article if scores on the posttest for the two groups were significantly different.

Middle grade students were also subjects in a study by Nelson (1967). The experimental group (n = 623) consisted of twelve fourth- and twelve sixth-grade classes, and the control group (n = 497) consisted of nine fourth- and nine sixth-grade classes. The experimental groups were taught a procedure for estimating answers to problems by the process of rounding numbers. The researcher conducted interviews with 30 students from each of the experimental and control groups. In the fourth and sixth grades the experimental groups performed significantly better on an estimation test developed by the researcher. In the sixth grade the experimental group performed significantly better than the control group on a test involving a knowledge of arithmetic concepts and applications. However, at the fourth grade, the control group scored significantly better than the experimental group in computation. As one would expect, the students in the experimental group with the highest IQs generally made the highest scores on the estimation tests.

In another experimental study by Ibe (1973), a group of sixth-grade students was taught, in a unit on angle measurement, to guess the measures of angles before measuring them with a protractor. The other group, a nonestimating group, always measured with a protractor. Using intelligence scores, the subjects from each of two classes had been randomly divided into two groups and each group then randomly assigned to a treatment. Prior to the experiment, significant correlations were found between estimating ability scores and four cognitive ability measures—general intelligence, number facility, spatial ability, and flexibility of perceptual closure. A criterion test was administered following the five-day experimental period. At the end of the experiment the groups taught to estimate had significantly higher scores than the control group not only on estimation but also on achievement and transfer.

In a somewhat different type of study, Sutherlin (1977) sought to determine if the use of calculators would improve estimation skills of fifth- and sixth-grade students. Subjects came from two classes from each of four Oregon schools. After random assignment to treatments, both classes were taught a unit on decimal estimation by the same teacher. The four experimental classes used calculators for precise calculations, and the four control classes used paper-and-pencil algorithms. Prior to the experiment there was no significant difference between the two groups on the estimation skill pretest. After the experiment both groups had gained in estimation skills, but there were still no significant differences between them on the posttest.

In another study that included calculators, Lawson (1978) randomly assigned seventh graders from one school to one of three treatment groups—a paper-and-pencil group, a calculator group, and an altered-calculator group. (Calculators were altered so that the four operation keys were not identifiable.) Four days before the experiment, there was a brief review of the desired estimation algorithm. A twenty-item pretest then measured students' ability to estimate answers to decimal problems. During the study students were to solve a sixteen-item computational task using the treatment assigned to their specific group. At the end of the study, students reestimated the twenty problems from the pretest, which had been reordered. Lawson concluded that the treatments had no effect on the estimation ability of the groups and, in addition, that computational ability correlated positively with ability to estimate. Analyses of videotapes made of a random sample of the altered-calculator group supported the conclusion that students did not use estimation processes to validate calculated answers.

Bestgen, Reys, Rybolt, and Wyatt (1980) studied 187 preservice elementary teachers who were enrolled in one of two mathematics preparatory courses or a methods-of-teaching-mathematics course. Students from seven sections were randomly assigned to two treatments, and students in two other sections became the control group. One experimental group was given weekly practice in estimation, and the second group was taught estimation strategies and had weekly practice. Not surprisingly, the groups receiving weekly practice had significantly greater gains in estimation than the control group. The group that received extra practice and instruction on strategies developed improved attitudes about mathematics, and these attitudes were better than those of the other groups. However, there were no significant differences in estimation performance between the two experimental groups. The researchers stated, "This research documents the value of providing preservice elementary teachers with computational estimation activities" (p. 135).

In an experimental study, Bright (1979) sought to determine if experienced teachers' skills in estimating linear relationships could be improved through practice. Subjects were 21 teachers of grades 6–8 enrolled in a summer workshop on the metric system. Skill in estimating was determined by the average percentage of error in each of three exercises in which subjects identified objects to match specified measurements. Bright concluded that these teachers' skill in estimating lengths seemed to improve with practice. There was no control group.

SUMMARY AND RECOMMENDATIONS

Little research has been devoted to the teaching and learning of estimation. As indicated, one reason involves the difficulty of testing estimation skills. It would be wise to continue attempts to develop and administer better instruments for assessing estimation abilities. More research also is needed in the choice of strategies. That is, once strategies are classified (and further study and replication is needed in this area), then how and why does a person choose a particular strategy? The most pressing need is for experimental research on estimation and how it should be taught.

Treatments in most of the experimental studies were short, lasting only a few days. Most of these studies offer encouraging but limited evidence that instruction and practice in estimation for students in the middle grades can result in improved performance on estimation exercises. More extensive studies need to be done in the middle grades. As for the early grades, even though estimation is not generally introduced this early, it would be helpful to determine if students in grades 1–3 can be effectively taught some simple estimation skills. Furthermore, since no experimental research was located at the secondary or college levels (with the exception of one study using preservice teachers), research seems to be needed at these two levels. At all levels, studies are needed that address the important interrelationship between estimation and conceptual understanding.

Finally, studies are needed that investigate how to train teachers in the area of estimation.

REFERENCES

Bestgen, Barbara J., Robert E. Reys, James F. Rybolt, and J. Wendell Wyatt. "Effectiveness of Systematic Instruction on Attitudes and Computational Estimation Skills of Preservice Elementary Teachers." *Journal for Research in Mathematics Education* 11 (March 1980): 124–36.

Bright, George W. "Measuring Experienced Teachers' Linear Estimation Skills at Two Levels of Abstraction." *School Science and Mathematics* 79 (February 1979): 161–64.

Carpenter, Thomas P., Terrence G. Coburn, Robert E. Reys, and James W. Wilson. "Notes from National Assessment: Estimation." *Arithmetic Teacher* 23 (April 1976): 296–302.

Carpenter, Thomas P., Mary Kay Corbitt, Henry S. Kepner, Jr., Mary Montgomery Lindquist, and Robert Reys. "Results of the Second NAEP Mathematics Assessment: Secondary School." *Mathematics Teacher* 73 (May 1980): 329–38.

Corle, Clyde G. "A Study of the Quantitative Values of Fifth and Sixth Grade Pupils." *Arithmetic Teacher* 7 (November 1960): 333–40.

———. "Estimates of Quantity by Elementary Teachers and College Juniors." *Arithmetic Teacher* 10 (October 1963): 347–53.

Faulk, Charles J. "How Well Do Pupils Estimate Answers?" *Arithmetic Teacher* 9 (December 1962): 436–40.

Freeman, Donald. *"The Fourth Grade Mathematics Curriculum as Inferred from Textbooks and Tests."* Paper presented at the meeting of the American Educational Research Association, April 1980, Boston. Mimeographed.

Hildreth, David J. "Estimation Strategy Uses in Length and Area Measurement Tasks by Fifth and Seventh Grade Students." (Doctoral dissertation, Ohio State University, 1980.) *Dissertation Abstracts International* 41 (1981): 4319A–4320A.

———. "The Use of Strategies in Estimating Measurements." *Arithmetic Teacher* 30 (January 1983): 50–56.

Ibe, Milagros D. "The Effects of Using Estimation in Learning a Unit of Sixth-Grade Mathematics." (Doctoral dissertation, University of Toronto, 1971.) *Dissertation Abstracts International* 33 (1973): 5036A.

Immers, Richard C. "Linear Estimation Ability and Strategy Use by Students in Grades Two through Five." (Doctoral dissertation, University of Michigan, 1983.) *Dissertation Abstracts International* 44 (1983): 416A.

Lawson, Thomas J. "A Study of the Calculator's and Altered Calculator's Effect upon Student Perception and Utilization of an Estimation Algorithm." (Doctoral dissertation, State University of New York at Buffalo, 1977.) *Dissertation Abstracts International* 39 (1978): 647A.

Levine, Deborah R. "Computational Estimation Ability and the Use of Estimation Strategies among College Students." (Doctoral dissertation, New York University, 1980.) *Dissertation Abstracts International* 41 (1981): 5013A.

National Council of Teachers of Mathematics. *An Agenda for Action: Recommendations for School Mathematics of the 1980s.* Reston, Va.: The Council, 1980.

Nelson, Nancy Z. "The Effect of the Teaching of Estimation on Arithmetic Achievement in the Fourth and Sixth Grades." (Doctoral dissertation, University of Pittsburgh, 1966.) *Dissertation Abstracts* 27 (1967): 4127A.

Paull, Duane R. "The Ability to Estimate in Mathematics." (Doctoral dissertation, Columbia University, 1971.) *Dissertation Abstracts International* 32 (1972): 3567A.

Reys, Robert E., and Barbara J. Bestgen. "Teaching and Assessing Computational Estimation Skills." *Elementary School Journal* 82 (November 1981): 116–27.

Reys, Robert E., James F. Rybolt, Barbara J. Bestgen, and J. Wendell Wyatt. "Processes Used by Good Computational Estimators." *Journal for Research in Mathematics Education* 13 (May 1982): 183–201.

Rubenstein, Rheta N. P. "Mathematical Variables Related to Computational Estimation." (Doctoral dissertation, Wayne State University, 1982.) *Dissertation Abstracts International* 44 (1983): 695A.

Schoen, Harold L., Charles D. Friesen, Joscelyn A. Jarrett, and Tonya D. Urbatsch. "Instruction in Estimating Solutions of Whole Number Computations." *Journal for Research in Mathematics Education* 12 (May 1981): 165–78.

Siegel, Alexander W., Lynn T. Goldsmith, and Camilla R. Madson. "Skill in Estimation Problems of Extent and Numerosity." *Journal for Research in Mathematics Education* 13 (May 1982): 211–32.

Stahlecker, Verlyn G. "A Study of Mathematics Preparation for Elementary School Teachers." (Doctoral dissertation, University of Montana, 1978.) *Dissertation Abstracts International* 40 (1979): 1330A.

Sutherlin, William N. "The Pocket Calculator: Its Effect on the Acquisition of Decimal Estimation Skills at Intermediate Grade Levels." (Doctoral dissertation, University of Oregon, 1976.) *Dissertation Abstracts International* 37 (1977): 5663A.

Suydam, Marilyn N. "Research on Mathematics Education Reported in 1981." *Journal for Research in Mathematics Education* 13 (July 1982): 241–318.

———. "Research on Mathematics Education Reported in 1982." *Journal for Research in Mathematics Education* 14 (July 1983): 227–93.

———. "Research on Mathematics Education Reported in 1983." *Journal for Research in Mathematics Education* 15 (July 1984): 243–315.

Suydam, Marilyn N., and J. F. Weaver. "Research on Mathematics Education Reported in 1979." *Journal for Research in Mathematics Education* 11 (July 1980): 241–320.

———. "Research on Mathematics Education Reported in 1980." *Journal for Research in Mathematics Education* 12 (July 1981): 241–319.

Swan, Malcolm D. "Experience, Key to Metric Unit Conversion." *Science Teacher* 37 (November 1970): 69–70.

Swan, Malcolm, and Orville Jones. "Distance, Weight, Height, Area, and Temperature Percepts of University Students." *Science Education* 55 (July/September 1971): 353–60.

Swan, Malcolm, and Orville E. Jones. "Comparison of Students' Percepts of Distance, Weight, Height, Area, and Temperature." *Science Education* 64 (July 1980): 297–307.